Atomic Design
Brad Frost

Copyright © 2016 Brad Frost All rights reserved

Publisher: Brad Frost
Copy editor: Owen Gregory
Print book designer: Rachel Arnold Sager
Ebook designer: Rachel Andrew
ISBN: 978-0-9982966-0-9

Proudly created in Pittsburgh, Pennsylvania

Table of Contents

1. Designing Systems
Create design systems, not pages — 8

2. Atomic Design Methodology
Atoms, molecules, organisms, templates, and pages — 40

3. Tools of the Trade
Pattern Lab and the qualities of effective style guides — 66

4. The Atomic Workflow
People, process, and making design systems happen — 96

5. Maintaining Design Systems
Making design systems stand the test of time — 144

Thanks & Acknowledgements — 184

Resources — 186

About the Author — 193

Foreword

It was 2013, and we huddled with Brad Frost and Jennifer Brook around a sunlit kitchen table in Brooklyn. The four of us had just begun work on a new website for TechCrunch, and we were sketching wireframes in Jennifer's apartment, wrestling with the new demands of responsive design. Brad pulled out his laptop: "I've been playing with a new idea."

Brad's screen looked like a webpage had exploded. Bits and pieces of UI floated free of each other, untethered by a unified design or hierarchy. It looked like a pile of spare parts from a web garage.

Brad flashed his crazy grin and nodded, "Great, right?" The three of us stared back blankly. Somebody coughed.

And then Brad Frost the front-end developer started talking like Brad Frost the chemist. He talked about atoms and molecules and organisms – about how large pieces of design can be broken down into smaller ones and even recombined into different large pieces. Instead of visualizing the finished recipe for the design, in other words, he was showing us the ingredients. And we lit up: this was a shift in perspective, a way to move away from conceiving a website design as a collection of static page templates, and instead as a dynamic system of adaptable components. It was an inspired way to approach this responsive website – and all responsive projects for that matter.

Brad's new idea was atomic design, and it changed the way we work in this astonishingly multi-device world. By thinking about interfaces simultaneously at both the large (page) level as well as the small (atomic) level, we streamlined our process: we introduced more rigorous thought into the role of every element; we fell into habits that improved the consistency of our UX; and crucially, we started working much faster and more collaboratively. Atomic design was our superpower.

In the early stages of the TechCrunch redesign, there was this moment where we talked about what we wanted the article page to be. Within an hour, Brad had a fully responsive version wired up from his kit of parts. That was the moment we realized just how quickly we were going to be able to move, a powerful case for investing in this clever, modular approach.

Almost four years later, we haven't looked back. Brad continued to refine his techniques and tools over the projects that followed, including blockbuster sites for Entertainment Weekly and Time, Inc. We've used these lessons to help in-house product teams make sites faster and with higher quality, build massive design systems for organizations looking to centralize their design and development work across international offices, and much more.

Atomic design gave us speed, creative freedom, and flexibility. It changed everything. We think it will do the same for you, too.

This wonderful book explains the philosophy, practice, and maintenance of atomic design systems. And it does so with the cheerful, helpful generosity that so describes Brad himself.

Brad's energy and big-hearted enthusiasm for the web and its makers are boundless. For years, Brad has worked at the forefront of responsive design technique – and he's shared everything along the way. His This Is Responsive site is the go-to resource for finding responsive solutions to any UX problem. His blog and Twitter feeds share his roadblocks and his solutions. When designers and developers follow Brad Frost, they get a fast and dense stream of practical, passionate insight for building beautiful and resilient websites. This book doubles down on all of that.

Given the chance, Brad would knock on the door of every designer and developer to personally deliver his message. We've watched with astonishment (and mild envy) as this whirling dervish has barnstormed around the globe to share his advice with hundreds of teams and organizations across six continents. (*Atomic design, coming soon to Antarctica!*) But even Brad Frost can't be everywhere at once, and we're delighted that he's detailed his ideas with such depth and good humor in this book.

Atomic design is blowing up around the world; it transformed our design practice; and we're excited for it to bring the same creative combustion to your process, too.

– Josh Clark[i] and Dan Mall[ii], Brad's frequent collaborators and his biggest fans

i https://bigmedium.com/
ii http://danielmall.com/

Chapter 1
Designing Systems

Create design systems, not pages

A long, long time ago, there were these things called *books*. Remember them? These contraptions were heavy and bulky and made from the pulp of dead trees. Inside these books were things called *pages*. You turned them, and they cut your fingers.

Awful things. I'm so glad these book things with their razor-sharp pages aren't around anymore.

Oh, wait...

Our paginated past

The page has been with us for a long time now. A few millennia, actually. The first books were thick slabs of clay created about 4,000 years ago, soon replaced by scrolls as the preferred way to consume the written word. And while reading technology has come a long way – from papyrus to parchment to paperback to pixels – the concept of the page holds strong to this day.

The page metaphor has been baked into the lexicon of the web since the very beginning. Tim Berners-Lee invented the World Wide Web so that he, his colleagues at CERN, and other academics could easily share and link together their world of *documents*. This document-based, academic genesis of the web is why the concept of the page is so deeply ingrained in the vocabulary of the internet.

So what?

As we'll discuss throughout this book, the way things are named very much impacts how they're perceived and utilized. Thinking of the web as pages has real ramifications on how people interact with web experiences, and influences how we go about creating web interfaces.

From the beginning, the page metaphor provided users with a familiar language with which to navigate this brave new World Wide Web. Concepts like bookmarking and pagination helped new web users explore and eventually master an entirely new medium using conventions they were already comfortable with.

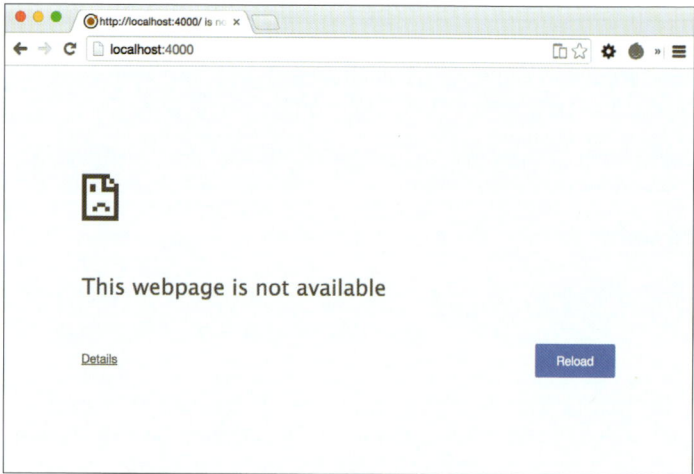

Chrome browser displaying 'This webpage is not available' message.

The page was – and continues to be – a very visible and helpful metaphor for the users of the web. It also has a profound influence on how web experiences are created.

In the early days of the web, companies looking to get online simply translated their printed materials onto their websites. But even though these brochure websites offered a very one-dimensional perspective of what the web could offer, viewing websites as digital representations of the printed page was easy for creators to wrap their heads around.

But we're now 25 years into this new medium, and this once necessary figure of speech has overstayed its welcome. Unfortunately, the **page metaphor continues to run deep with respect to how we scope and execute our web projects.** Here are just a few examples I hear on a regular basis:

"We're a startup looking to launch a five-*page* website this October..."

"Brad, how long will the home*page* take to build?"

"How are we ever going to redesign this university website that contains over 30,000 *pages*?!"

All of the statements above make the fundamental mistake of assuming a page is a uniform, isolated, quantifiable thing. The reality is that the web is a fluid, interactive, interdependent medium. As soon as we come to terms with this fact, the notion of the page quickly erodes as a useful means to scope and create web experiences.

How long will a homepage take to build? Well, that sort of depends on what's on it, right? Maybe the homepage simply consists of a tagline and a background image, which means it could be done by lunch. Or maybe it's chock-full of carousels, dynamic forms, and third-party integrations. In that case, maybe the homepage will take several months to complete.

As for the 30,000-page university website, it might be tempting to declare, "Thousands of pages?! Wow, that sounds challenging!" But in reality, those 30,000 pages may consist of three content types and two overarching layouts.

Ultimately, a project's level of effort is much better determined by the functionality and components[1] contained within those pages, rather than on the quantity of pages themselves.

The page metaphor has served its purpose helping users familiarize themselves with the web, and provided creators with the necessary transitional language with which to create for a brand new medium. But to build thoughtful interfaces meant to be served to a multitude of connected devices, the time has come for us to evolve beyond the page.

Tearing up the page

Thankfully, the web community is hard at work establishing principles and practices to help us effectively talk about and create for the web. And there's one concept that keeps popping up in every conversation about how to make successful web experiences: **modularity**.

Modularity predates the web by a long shot. The Industrial Revolution brought about interchangeable parts and Henry Ford's assembly line forever transformed the automobile manufacturing

---

[1] http://bradfrost.com/blog/post/scope-components-not-pages/

process. The earliest cars and components were individually crafted, which led to many safety and maintainability nightmares. Ford broke the automobile down into its component parts and modularized the assembly process. The results spoke for themselves: more uniform, more reliable, safer cars rolled out of the factory, and in record time to boot.

As the machine age became the computer age, computer scientists began practicing object-oriented programming and establishing important modular concepts like *separation of concerns* and the *single responsibility principle*. It is from this world that the World Wide Web was born, so it's no surprise that modular design[2] quickly became a design principle for the architecture of the web.

Slowly, but surely, these concepts found their way into web designers' workflows. In the early 2000s we saw the introduction of libraries like YUI[3] and jQuery UI[4] that provided developers with a toolkit of widgets and patterns to create more consistent user interfaces.

If modularity has been around for such a long time, why are we talking about it now?

The short answer is that modularity matters more than ever. Right now, our entire industry is drowning in a sea of devices, viewport sizes, and online environments. And things aren't slowing down anytime soon.

> *Disruption will only accelerate. The quantity and diversity of connected devices – many of which we haven't imagined yet – will explode, as will the quantity and diversity of the people around the world who use them. Our existing standards, workflows, and infrastructure won't hold up. Today's onslaught of devices is already pushing them to the breaking point. They can't withstand what's ahead.*
>
> – The Future-Friendly manifesto[5]

---

2 http://www.w3.org/DesignIssues/Principles.html#Modular
3 http://yuilibrary.com/
4 http://jqueryui.com/
5 http://futurefriendlyweb.com/

These are just some of the connected devices we need to worry about.

Like it or not, this multi-device universe is our reality. It was hard enough to get our web pages to display consistently in a handful of desktop browsers, but we're now tasked with ensuring our web experiences look and function beautifully on a dizzying array of smartphones, tablets, phablets, netbooks, notebooks, desktops, TVs, game consoles, and more.

To address this reality while maintaining our sanity, it's absolutely necessary for us to take a step back and break these giant responsibilities into smaller, more manageable chunks.

And that's exactly what folks are doing. The spirit of modularity is weaving its way into every aspect of the web creation process and having profound effects on organizations' strategy, process, content, design, and development.

A manageable strategy

Every organization is finally realizing that bulldozing their entire website and replacing it with a New-And-Shiny™ website every three to eight years isn't (and never was) an optimal solution.

Out with the old! In with the new! It's certainly an attractive prospect. But even before the launch party confetti is swept up, the calls start coming in. "You moved my cheese!" cry the users, who spent years learning the previous interface and functionality.

When massive redesigns launch with significant changes to the experience, users get knocked down what Jared Spool calls the "Magic Escalator of Acquired Knowledge"[6]. Huge redesigns are a jolt to the system, and newly frustrated users have to spend a great deal of time and energy relearning the experience in order to slowly climb back up that escalator of acquired knowledge.

In addition to disorienting users, these monolithic redesigns don't get to the organizational root of the problem. Without a fundamental change in process, history is bound to repeat itself, and what's New-And-Shiny™ today becomes Old-And-Crusty™ tomorrow. The cycle repeats itself as companies push off minor updates until the next big redesign, ultimately paralyzing themselves and frustrating users in the process.

Thankfully, even massive organizations are taking cues from the smaller, leaner startup world and striving to get things out the door quicker. By creating *minimum viable products* and shipping often to iteratively improve the experience, organizations are able to better address user feedback and keep up with the ever-shifting web landscape.

Moving away from Ron Popeil-esque, set-it-and-forget-it redesigns requires deliberate changes in organizational structure and workflow. Which is a heck of a lot easier said than done.

An iterative process

If I had a quarter for every time I heard some stakeholder declare "We're trying to be more agile," I'd be orbiting the earth in my private spacecraft instead of writing this book.

Wanting to be more agile is commendable. But *agile* is a loaded term, with big differences between capital-A Agile and lowercase-a agile. Capital-A Agile is a specific methodology for software

[6] http://www.uie.com/articles/magic_escalator/

development, equipped with a manifesto[7] and accompanying frameworks like Scrum[8] and Lean[9].

Lowercase-a agile is more of an informal desire to create an efficient process. This desire may certainly involve adopting general principles[10] from capital-A Agile, but it may not involve adopting the Agile process in its entirety. Project manager Brett Harned explains:

> We want to be more agile; we're embracing change, continuing improvement, being as flexible as possible, and adapting as we see fit. The thing is, we won't ever truly be Agile, as the Manifesto states. That's okay, as long as we say what we will be.
>
> - Brett Harned[11]

Organizational structure, client relations, personalities, and so on all play major roles in determining a project's process. The trick is to find the process that works best for you, your organizational constraints and opportunities.

Even though it may be impossible to adopt a truly Agile process, it's still a sound idea to work in cross-disciplinary teams, get into the final environment faster, ship early and often, and break bigger tasks into smaller components. In chapter 4, we'll detail how to establish an effective pattern-based workflow.

Modularizing content: I'm on Team Chunk

> Get your content ready to go anywhere, because it's going to go everywhere.
>
> - For A Future-Friendly Web[12]

7   http://www.agilemanifesto.org/
8   http://en.wikipedia.org/wiki/Scrum_%28software_development%29
9   http://en.wikipedia.org/wiki/Lean_software_development
10 http://www.agilemanifesto.org/principles.html
11 http://cognition.happycog.com/article/diy-process
12 http://bradfrost.com/blog/web/for-a-future-friendly-web/

Publishing content for the Web used to be a fairly straightforward endeavor, as the desktop web was the only game in town. Oh, how things have changed. Today, our content is consumed by a whole slew of smartphones, dumb phones, netbooks, notebooks, tablets, e-readers, smartwatches, TVs, game consoles, digital signage, car dashboards, and more.

To properly address this increasingly diverse and eclectic digital landscape, we need to dramatically overhaul our perception of content and the tools we use to manage it.

> In the future, what I believe is that we are going to have better content management and content publishing tools. We are going to have ways to take well-structured content, well-designed chunks of content that we can then figure out how we want to restructure and publish and display in a way that's going to be right for the appropriate platform.
>
> - Karen McGrane[13]

Thankfully, this future is starting to take shape. Organizations are recognizing the need to create modularized content to better reach their audience wherever they may be. And content management systems are evolving beyond their web publishing platform roots into tools that can elegantly create and maintain modular content. While sophisticated content management systems have existed for years in the form of custom solutions like NPR's COPE (Create Once, Publish Everywhere) platform[14], smart modular thinking is making its way into mainstream content management systems.

Classy code

Modularity has long been a staple principle in the world of computer science, as we discussed earlier. While this principle existed long before the web was invented, it has taken some time for modularity to become engrained in the minds and hearts of web developers.

---

13 http://karenmcgrane.com/2012/09/04/adapting-ourselves-to-adaptive-content-video-slides-and-transcript-oh-my/

14 http://www.programmableweb.com/news/cope-create-once-publish-everywhere/2009/10/13

Despite being around since 1995, JavaScript, the programming language of the web, first had to endure some growing pains to mature into the capable, respected language it is today. Now that JavaScript has grown up, developers can apply those tried-and-true computer science principles to their web development workflows. As a result, we're seeing folks develop sophisticated JavaScript patterns[15] and architectures.

Applying modular programming principles to JavaScript is a bit of a no-brainer, since JavaScript is itself a programming language. But object-oriented thinking is weaving its way into other aspects of the web as well, including CSS, the styling language of the web. Methodologies like OOCSS[16], SMACSS[17], and BEM[18] have cropped up to help web designers create and maintain modular CSS architectures.

Visually repaired

Not only is modularity infiltrating the code side of style on the web, it's revolutionizing how visual designers approach modern web design.

As the number of viewports and environments proliferate, it's become untenable to produce static mockups of every page of a web experience. As Stephen Hay quipped, presenting fully baked Photoshop comps "is the most effective way to show your clients what their website will never look like."

That's not to say static design tools like Photoshop and Sketch aren't important. Far from it. But it's the way we use these tools that has changed dramatically. While creating hundreds of full-on comps isn't realistic, these static tools excel at providing a playground to establish what Andy Clarke calls "design atmosphere":

---

15 http://addyosmani.com/resources/essentialjsdesignpatterns/book/
16 http://oocss.org/
17 https://smacss.com/
18 http://csswizardry.com/2013/01/mindbemding-getting-your-head-round-bem-syntax

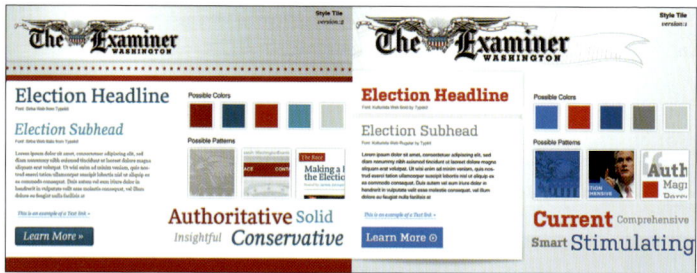

Style tiles, a concept created by designer Samantha Warren, allow designers to explore color, typography, and texture without having to develop fully realized comps.

> Atmosphere describes the feelings we get that are evoked by colour, texture and typography. You might already think of atmosphere in different terms. You might call it "feel", "mood" or even "visual identity." Whatever words you choose, the atmosphere of a design doesn't depend on layout. It's independent of arrangement and visual placement. It will be seen, or felt, at every screen size and on every device.
>
> - Andy Clarke[19]

Establishing design atmosphere early is critical to a project's success, which is why designers have found ways to facilitate these important conversations without having to generate full mockups. Designer Samantha Warren developed design artifacts called style tiles[20], which demonstrate color, type, and texture explorations in a nice encapsulated one-pager. Designer Dan Mall built on Samantha's idea with a concept called element collages[21], which demonstrate design atmosphere explorations in an exploded collage of interface elements.

By breaking visual explorations into smaller chunks, designers save time and effort while avoiding presenting unrealistic, premature layouts to clients. More importantly, these approaches

19 http://stuffandnonsense.co.uk/blog/about/an-extract-from-designing-atoms-and-elements
20 http://styletil.es/
21 http://danielmall.com/articles/rif-element-collages/

shift stakeholders away from simply reacting to a pretty picture, and instead facilitate crucial conversations about overall design direction and how they relate to the project's goals. We'll discuss these concepts in more detail in chapter 4, but suffice it to say the visual design workflow is changing in a big way!

Systematic UI design

> We're not designing pages, we're designing systems of components.
>
> - Stephen Hay[22]

What is an interface made of? What are our Lego bricks? What are our Subway sandwich pieces that we combine into millions of delicious combinations? It's these questions that we've been asking ourselves more and more now that we're sending our interfaces to more and more places.

A few years ago Ethan Marcotte introduced us to the idea of responsive web design[23] and its three core tenets: fluid grids, flexible media, and CSS media queries. These three ingredients provided a much-needed foundation for designers to create flexible layouts that smartly adapt to any screen size. Perhaps more importantly, responsive design helped get designers excited about creating thoughtful, adaptable, multi-device web experiences.

As designers quickly discovered, creating multi-device web experiences involves a lot more than creating squishy pages. Each individual piece of an interface contains its own unique challenges and opportunities in order for it to look and function beautifully across many screen sizes and environments.

How can we present primary navigation – typically displayed as a horizontal list on large screens – in a thoughtful way on smaller screens? How do lightboxes, breadcrumbs, and carousels translate to smaller viewports and alternate input types? It's these questions

---

22 http://bradfrost.com/blog/mobile/bdconf-stephen-hay-presents-responsive-design-workflow/
23 http://alistapart.com/article/responsive-web-design

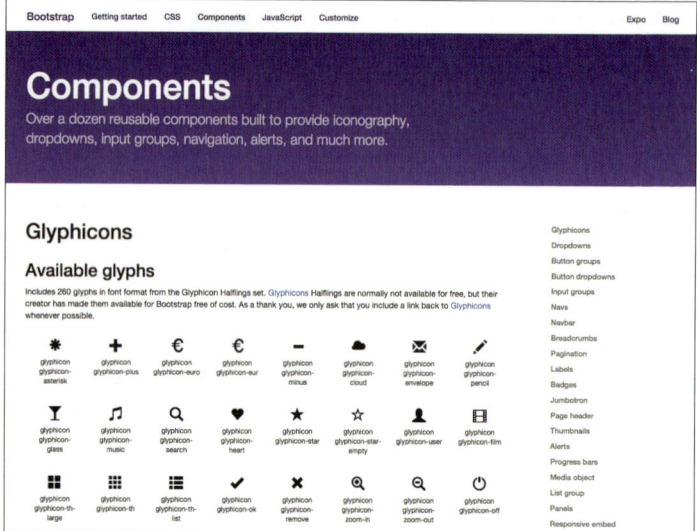

Bootstrap provides a collection of UI components to speed up development.

that led me to create This Is Responsive[24], a showcase of responsive patterns that demonstrate the various ways a particular component could be executed in a responsive environment.

While This Is Responsive is successful at articulating how these interface patterns can scale across screen sizes and environments, it's still up to designers and developers to put these patterns into action. And as it turns out, that's a lot of work.

UI frameworks, in theory and in practice

Designers and developers are already strapped for time and resources, and they're now being tasked with making interfaces that look and function beautifully in any environment. That's a very tall order.

This need to address growing device diversity while still sanely getting projects out the door has given rise to front-end

---

24   http://bradfrost.github.io/this-is-responsive/index.html

frameworks like Foundation by Zurb[25] and Bootstrap[26]. These user interface frameworks provide designers with a collection of preassembled HTML patterns, CSS styles, and JavaScript to add functionality to interactive components like dropdowns and carousels. In essence, these frameworks are handy tool kits for quickly assembling interfaces.

And boy are these things popular. As I'm writing this, Bootstrap is the most popular repository on the code-sharing site GitHub[27], with over 77,000 stars and 30,000 forks. These frameworks' popularity is a testament to the fact that designers and developers are seeking solid ground to stand on in this ever-complex web landscape.

One of the most attractive aspects of these frameworks is *speed*. Frameworks like Bootstrap allow designers to get ideas off the ground quickly, rapidly create prototypes, and launch sites sooner. Because the patterns provided by a tool kit are already cross-browser tested, developers can spend their time on more important tasks rather than beating their heads against a table testing some archaic version of Internet Explorer. And in case designers do get stuck, these frameworks' communities can provide helpful support and advice.

For freelancers, this increase in speed might mean they can take on an extra project or three, yielding more financial stability for the year. And in the startup world – a place where Bootstrap is omnipresent – minimum viable products can launch sooner, leading to faster answers regarding the products' viability.

So frameworks like Bootstrap are insanely popular design systems that provide well-tested components, resulting in consistent designs and faster launches. What's not to love? Well, like most everything in life, there are cons right there alongside the pros.

Trouble in framework paradise

When I was a kid, I'd watch sci-fi movies and TV shows with a strange fascination. There was one question I could never quite shake: *why are they all dressed the same?*

---

25 http://foundation.zurb.com/
26 http://getbootstrap.com/
27 https://github.com/

I could only guess that given enough time, we *solve fashion*. "Say, these jumpsuits are pretty snazzy, and comfortable too! Let's just all wear these from now on." "Sounds good to me!"

In the future, everyone dresses the same. Illustration credit: Melissa Frost.

Of course, that's not how human beings work. We all have different tastes, goals, and desires. Variety, as they say, is the spice of life, and fashion, music, and design reflect our diverse nature. Yet on the web we tend to fall into the trap of wanting everyone to do things the same way. "Why don't all browsers just standardize on WebKit?" "Why can't device manufacturers just use the same screen sizes?" "Always use jQuery!" "Never use jQuery!" "Just use frameworks!" "Never use frameworks!"

Just like the real world, the diverse needs, goals, and desires of web projects lead to a myriad of different solutions. Of course, there's a time and place for everything, and designers and developers need the discernment to know which tools to use and when.

Front-end frameworks are tools that provide a specific solution and a particular look and feel. While those solutions help speed up development, the resulting experiences end up resembling those sci-fi jumpsuits. When everyone uses the same buttons, grids, dropdowns, and components, things naturally start to look the same. If Nike, Adidas, Puma, and Reebok were to redesign their respective sites using Bootstrap, they would look substantially similar. That's certainly not what these brands are going for. Sure, each brand can modify and extend the default look and feel, but after a while customization means fighting the framework's given structure, style, and functionality.

In addition to look-alike issues, these frameworks can add unnecessary bloat to an experience. It's fantastic that frameworks provide plenty of prebuilt components and functionality, but a large percentage of designers and developers won't adopt every aspect of the framework. Unfortunately, users still have to download the framework's unused CSS and JavaScript, resulting in slower page loads and frustration.

On the flip side of that coin, frameworks might not go far enough, leading to developers needing to create a substantial amount of custom code to achieve their projects' goals. At some point, a threshold is crossed where the initial benefits of using a framework – namely development – are outweighed by the time spent modifying, extending, and fixing the framework.

And then there's the issue with naming. Using a framework means subscribing to someone else's structure, naming, and style conventions. Of course, it's important to establish a useful front-end lexicon, but what makes sense for an organization might not be what comes out of a framework's box. I, for one, would balk at the idea of using Bootstrap's default component for a featured content area they call a "jumbotron". How a framework's naming conventions jive with an existing codebase and workflow should be properly discussed before jumping on board the framework train.

Now that we've put frameworks through the wringer, it's important to take a step back and recognize that conceptually these frameworks are very much on point. It's an excellent idea to work with a design tool kit that promotes consistency and speeds up development time. While discussing the redesign of Microsoft's homepage by Austin-based web shop Paravel, developer Dave Rupert stressed the importance of creating and delivering a design system to their client. Dave wonderfully articulated that it's not necessarily about using Bootstrap for every client, but rather creating "tiny Bootstraps for every client."

> Responsive deliverables should look a lot like fully-functioning Twitter Bootstrap-style systems custom tailored for your clients' needs. These living code samples are self-documenting style guides that extend to accommodate a client's needs as well as the needs of the ever-evolving multi-device web.
>
> – Dave Rupert[28]

It's not just about using a design system, it's about creating *your* system.

Design systems save the day

So what do robust design systems look like? What form do they take? How do you create, maintain, and enforce them?

The cornerstones of good design systems are *style guides*, which document and organize design materials while providing guidelines, usage, and guardrails.

As it happens, there are many flavors of style guides[29], including documentation for brand identity, writing, voice and tone, code, design language, and user interface patterns. This book won't detail every category of style guide, but it's important to take a look at each to better understand how each style guide influences the others, and how style guides for the web fit into a larger ecosystem.

Brand identity

Brand identity guidelines define the assets and materials that make a company unique. Logos, typography, color palettes, messaging (such as mission statements and taglines), collateral (such as business card and PowerPoint templates), and more are aggregated and described in brand identity guidelines.

It's essential for a brand to present itself in a cohesive manner across an increasing number of media, channels, and touchpoints. How can everyone within an organization speak in one voice and

---

28 http://daverupert.com/2013/04/responsive-deliverables/
29 http://bradfrost.com/blog/post/style-guides/2

West Virginia University's brand style guide.

feel part of a singular entity? How do third parties know which Pantone colors to use and how to correctly use the brand's logo? Brand identity guidelines provide answers to these fundamental questions in one centralized hub.

Historically, brand identity guidelines were contained in hard-cover books (remember, those things with the pages?), but as with everything else, brand style guides are making their way online.

Design language

While brand identity guidelines are fairly tactile, design language guidelines are a bit harder to pin down. Design language style guides articulate a general design direction, philosophy, and approach to specific projects or products.

To present itself in a cohesive way across a growing range of products and media, Google developed a design language called *material design*. The material design style guide[30] defines its overarching design philosophy, goals, and general principles, while also providing specific applications of the material design language.

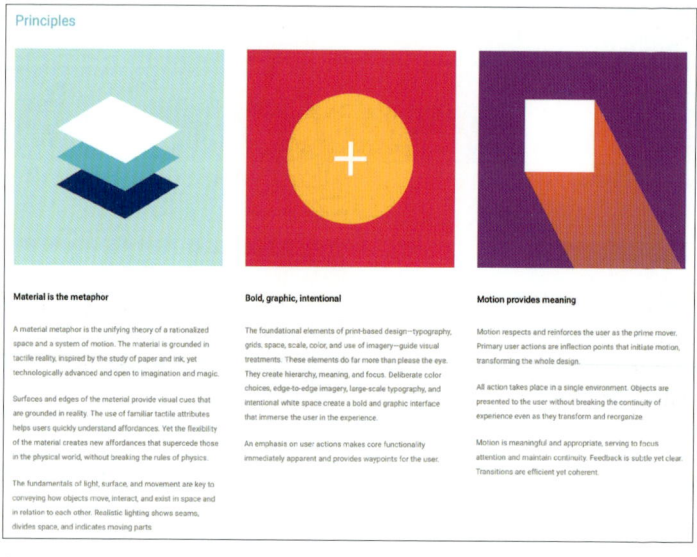

Google's material design language.

Design language style guides can (and usually do) incorporate aspects of other style guide categories in order to make high-level concepts a bit more tangible.

Design language guidelines aren't set in stone the way brand guidelines are. For example, one day Google will likely develop a new design language to replace material design, so while Google's overall brand will remain intact, the design vocabulary around its products will change.

---

30 http://www.google.com/design/spec/material-design/introduction.html

Voice and tone

People interact with brands across a huge array of channels and media. In addition to the digital media we've discussed so far, brands also operate in print, retail, outdoor, radio, TV, and other channels. When a brand must communicate across so many varied touchpoints, speaking in a unified, consistent manner becomes critical to a brand's success.

> *A brand's voice stays the same from day to day, but its tone has to change all the time, depending on both the situation and the reader's feelings.*
>
> – Kate Kiefer Lee[31]

Voice is an elemental aspect of a brand's identity, so typically brand identity guidelines include some reference to the brand's voice. However, these guidelines usually aren't very nuanced, which is why voice and tone guidelines are so important.

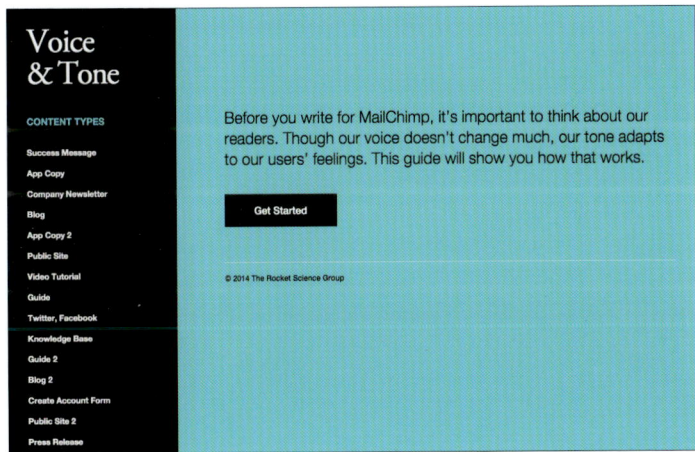

MailChimp's Voice and Tone guidelines

---

[31] http://www.slideshare.net/katekiefer/kkl-c-sforum

Voice and tone guidelines get into the weeds by articulating how the company's voice and tone should shift across a variety of scenarios. MailChimp's brilliant voice and tone guidelines[32] define how the brand's tone changes across content types, so that when a user's credit card is declined, writers know to shift away from their generally cheeky and playful tone of voice and adopt a more serious tone instead.

Writing

The rise of the web and content-managed websites makes it easier than ever for many people within an organization to publish content. This, of course, can be a double-edged sword, as maintaining a consistent writing style for an organization with many voices can be challenging. Writing style guides provide every author some guidelines and guardrails for contributing content.

Writing style guides can be extremely granular, defining particulars around punctuation and grammar, but they don't always have to be so detailed. Dalhousie University's writing style guide[33] provides a concise list of principles and best practices for content contributors to follow.

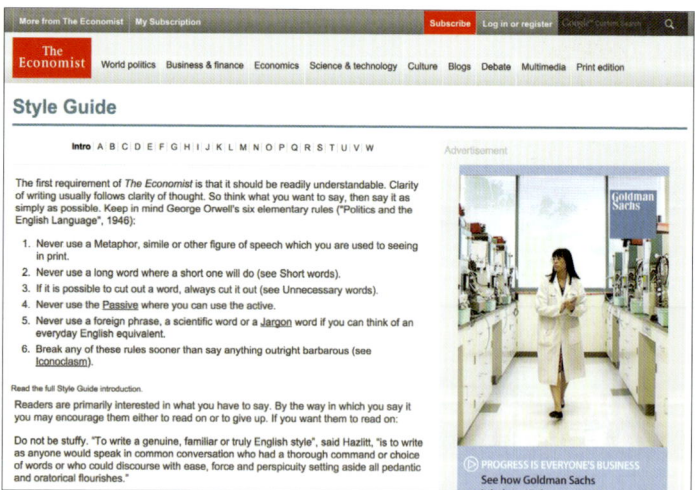

The Economist's writing style guide.

32 http://voiceandtone.com/

33 http://www.dal.ca/webteam/web_style_guide/writing_for_the_web.html

Code style guides

It's essential for teams to write legible, scalable, maintainable code. But without a way to promote and enforce code consistency, it's easy for things to fall apart and leave every developer to fend for themselves.

Code style guides provide conventions, patterns, and examples for how teams should approach their code. These guidelines and guardrails help rein in the madness so that teams can focus on producing great work together rather than refactoring a bunch of sloppy, inconsistent code.

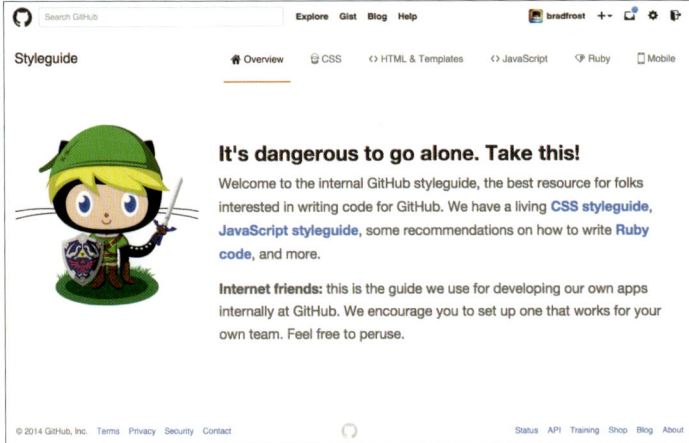

GitHub's code style guide provides best practices for writing HTML, CSS, JavaScript, and Ruby within their organization.

Pattern Libraries

And now for the main event. Pattern libraries, also known as front-end style guides, UI libraries, or component libraries, are quickly becoming a cornerstone of modern interface design.

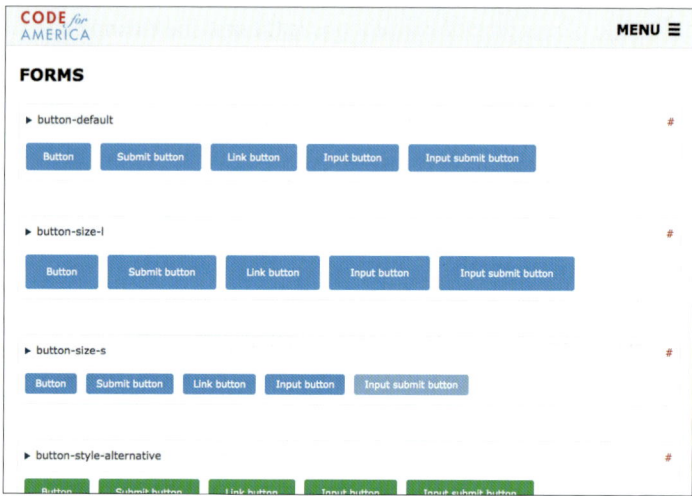

Code for America's pattern library

The rest of this book will concentrate on how to approach interface design in a systematic manner, and detail how to establish and maintain pattern libraries.

Style guide benefits

Getting UIs to work across a myriad of screen sizes, devices, browsers, and environments is a tall order in and of itself. But once you factor in other team members, clients, stakeholders, and organizational quirks, things start looking downright *intimidating*.

Style guides are important tools that help prevent chaos, both from a design and development standpoint and also from an organizational perspective. Here's why style guides are now essential tools for modern web design and development.

Consistently awesome

Web style guides promote consistency and cohesion across a user interface. This consistency benefits both the people who create these interfaces and also their users.

I recently visited my health insurance provider's website to pay my bill. In the course of five clicks, I was hit with four distinct interface designs, some of which looked like they were last touched in 1999. This inconsistent experience put the burden on me, the user, to figure out what went where and how to interpret disparate interface elements. By the time I got to the payment form, I felt like I couldn't trust the company to successfully and securely process my payment.

Style guides help iron out these inconsistencies by encouraging reuse of interface elements. Designers and developers can refer back to existing patterns to ensure the work they're producing is consistent with what's already been established.

Even third parties responsible for matching their UIs with the look and feel of a company's internal UIs can make great use of a style guide. Externally hosted experiences like payment portals or localized sites can better match the look and feel of the primary experience by applying the styles defined in the guide.

Making style guides central to your process results in user interfaces that feel more united and trustworthy, which helps users accomplish their tasks faster and empowers them to master the interface.

A shared vocabulary

What does "utility toolbar" mean? Does everyone understand what a "touch slider hero" is?

As the number of people working on a project increases, it becomes all too easy for communication breakdowns to occur. It's not uncommon for different disciplines to have different names for the same module, and for individuals to go rogue and invent their own naming conventions. For true collaboration to occur, it's essential for teams to speak a common language. Style guides are there to help establish that shared vocabulary.

Giving names to patterns like 'Blocks Three-Up' in Starbucks' style guide helps team members speak the same language.

Style guides establish a consistent, shared vocabulary between everyone involved in a project, encouraging collaboration between disciplines and reducing communication breakdowns.

Education

In her book *Front-End Style Guides*[34], Anna Debenham deftly explains the many advantages of creating style guides, including one of the most crucial benefits: education.

> *Education is as important as documentation. A style guide can show clients that websites are systems rather than collections of pages.*
>
> – Anna Debenham[35]

Style guides demonstrate to clients, stakeholders, and other disciplines that there's a lot of really thoughtful work going into a website's design and development beyond just "Hey, let's make a new website." **A pattern library communicates the design language in a very tangible way**, which helps stakeholders understand that an underlying system is determining the final interface.

Style guides can help alleviate what I call *special snowflake syndrome*, where certain departments in an organization think that they have unique problems and therefore demand unique solutions. By exposing the design system in the form of a style guide, these special snowflakes can better appreciate consistency and understand why their requests for custom designs receive pushback.

34, 35   http://maban.co.uk/projects/front-end-style-guides/

An empathetic workflow

Education isn't just important for clients and stakeholders. A good style guide helps inform designers and developers of the tools they have in their toolbox, and provides rules and best practices for how to use them properly.

By making a style guide a cornerstone of your workflow (which we'll detail in chapter 4), **designers and developers are forced to think about how their decisions affect the broader design system.** It becomes harder to go rogue and easier to think of the greater good. And this is exactly where you want team members to be.

A style guide provides a home for each discipline to contribute their respective considerations and concerns for patterns. By collecting all these considerations under one roof, the style guide becomes a hub for everyone involved in the project, which helps each discipline better understand the design system from many perspectives.

Testing, testing, 1-2-3

The homepage is broken, you say? Well, what exactly is breaking it?

The ability to pull an interface apart into its component pieces makes testing a lot easier. A style guide allows you to view interface patterns in isolation, allowing developers to zero in on what's causing errors, browser inconsistencies, or performance issues.

Speed

Earlier in the chapter we discussed how *faster design and development* is one of the main reasons why UI frameworks like Bootstrap are so popular. We're under pressure to get projects out the door as soon as humanly possible. By developing your own design system, you can reap those same speed rewards as the out-of-the-box UI tool kits.

It's true that devising an interface design system and creating a custom pattern library initially takes a lot of time, thought, and effort. But once the pattern library is established, subsequent design and development becomes much faster, which tends to make everybody happy.

Federico Holgado, lead UX developer at MailChimp, explained[36] how MailChimp's pattern library initially consisted of patterns created from the four primary screens of their app. But once they moved on to other areas of the site, they realized they were able to use existing patterns rather than having to generate brand new patterns from scratch every time.

> ...Once we did that, as we were implementing things in other pages we started to realize: man, this system will actually work here and this system will actually work here and here.
>
> – Federico Holgado[37]

In it for the long haul

There's no doubt style guides help teams effectively get things done in the here and now. But much like a fine wine, style guides increase in value over time. The beautiful thing about interface design systems is that they can and should be modified, extended, and refined for years to come.

As previously mentioned, creating a custom pattern library requires a lot of hard work up front, but that hard work should provide a structural foundation for future iteration and refinement. Lessons learned from analytics, user testing, A/B testing, and experience should be incorporated into the style guide, making it a powerful hub for truth, knowledge, and best practices.

Better yet, even if you were to undertake a major redesign you'll find that many of the structural interface building blocks will remain the same. You'll still have forms, buttons, headings, other common interface patterns, so there's no need to throw the baby out with the bath water. A style guide provides a rock-solid foundation for all future work, even if that future work may look totally different.

36, 37   http://styleguides.io/podcast/federico-holgado/

Style guide challenges

By now the benefits of creating design systems should be abundantly clear, and hopefully visions of sugar plums and beautiful style guides are dancing through your head. But to reach style guide nirvana, you must first overcome the many treacherous challenges that come with the territory.

The hard sell

To benefit from style guides, organizations must first appropriate the necessary time and budget to make them happen. **That requires organizations to overcome the short-term mentality that all too often creeps its way into company culture.**

The long-term benefits that style guides provide are obvious to those who are already thinking about the long game. The challenge, therefore, becomes convincing those stuck in a short-term, quarter-by-quarter mindset that establishing a thoughtful design system is a smart investment in the future.

A matter of time

> *The hard part is building the machine that builds the product.*
>
> – Dennis Crowley[38]

Perhaps the biggest, most unavoidable challenge is that **style guides are time-consuming to create**. I don't know about you, but I don't go into work every day twiddling my thumbs wondering what to do with my time. I've never met a person who isn't feeling pressure to get work out the door, and this pressure naturally leads to focusing on the primary web project. Unfortunately, aggressive timelines and finite budgets detract from the effort required to make style guides happen, even when teams are committed to the cause.

---

38 http://techcrunch.com/2011/03/03/founder-stories-foursquare-crowley-machine/

Auxiliary Projects

Pattern libraries are often treated as auxiliary projects, rather than as the component parts of the final product. By treating pattern libraries as something separate from the core project, they tend to fall into the *nice to have* category and become first on the chopping block when the going gets tough.

This auxiliary project conundrum reminds me of sentiments I often hear around factoring accessibility into projects. They say, "Oh, we wish we had the time and budget for accessibility, but…" The notion that accessibility (and other principles like performance and responsiveness) is a costly extra line item is a fallacy. Pattern libraries, like accessibility, are good ideas to bake into your workflow whether or not the project plan explicitly calls for them.

Maintenance and governance

Even if time and money are allocated to establish style guides, these valuable tools often die on the vine if they're not given the focus they need to reach their true potential.

A maintenance and governance strategy is critical to style guides' success. Style guides will be thrown in the trash (right beside all those PSDs and wireframes) and abandoned without a proper strategy in place for who will manage, maintain, and enforce them.

Style guide maintenance is a hugely important topic and deserves to be covered in detail, so we'll dive into how to create maintainable style guides in chapter 5.

Audience confusion

Style guides can be misunderstood as tools useful only to designers or developers, which leads to a lack of visibility that immediately limits their effectiveness. Instead of serving as a watering hole for everyone in the organization, a style guide can become a best-kept secret guarded by one discipline. Color me naive, but I don't think this helps foster a culture of collaboration.

Without thinking of broader audiences, style guides may come across as too vague or too technical, which can intimidate other disciplines and lead them to believe these resources aren't for them.

Style guide structure

For style guides to be useful resources for everyone in an organization, they should clearly convey what they are and why they matter. Style guides should be attractive, inviting, visible, clear, and easy to use. As mentioned above, they should be aware that a whole host of audiences will be viewing them, so should therefore aim to be welcoming and useful for as many people as possible.

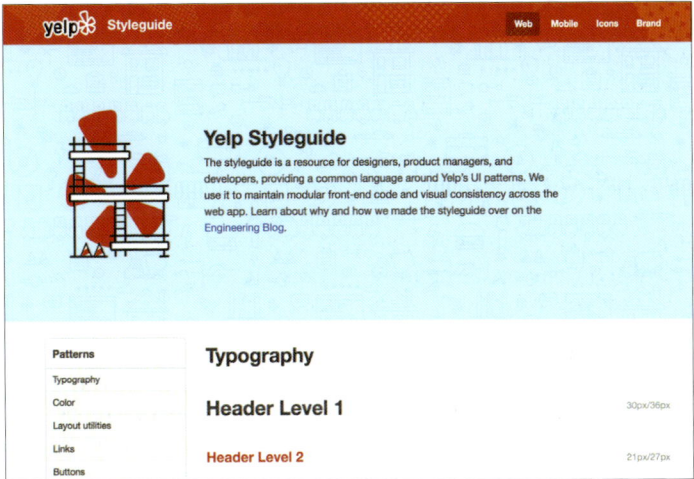

Yelp's style guide homepage sports a handsome design and important intro text explaining the purpose and audience for the guide.

Lack of context

Context is key to understanding a design system. Unfortunately, most pattern libraries out in the wild don't provide any hints as to when, how, and where their components get used. Without providing context, designers and developers don't know how global a particular pattern is, and as a result wouldn't know which pages of their app would need to be revisited, QA'd, and tested if changes were made.

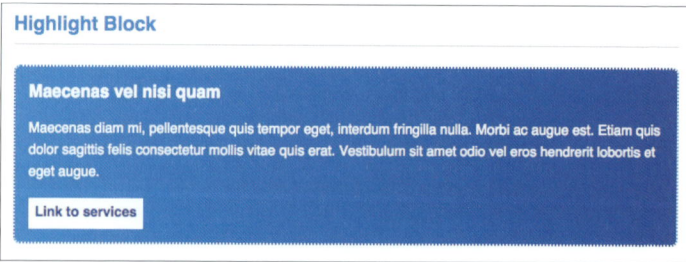

'Highlight Block' looks useful, but where is this pattern being used?

Lacking a clear methodology

As much as I adore the pattern libraries out there[39], I can't help but notice a lack of structure in many of them. Don't get me wrong, I think it's absolutely fantastic that teams are thinking systematically and are documenting their UI patterns. But I often feel like many pattern libraries are little more than loosely arranged sprays of modules. I think there's room for improvement.

In search of an interface design methodology

For us to create experiences for this eclectic web landscape, we must evolve beyond the page metaphor that's been with us since the birth of the web. Thankfully, organizations are embracing modularity across every aspect of the web creation process, which is leading to smarter work and more sustainable systems.

As the number of devices, browsers, and environments continues to increase at a staggering rate, the need to create thoughtful, deliberate interface design systems is becoming more apparent than ever.

Enter atomic design.

---

39 http://styleguides.io/examples.html

Chapter 2
Atomic Design Methodology

Atoms, molecules, organisms, templates, and pages

My search for a methodology to craft interface design systems led me to look for inspiration in other fields and industries. Given this amazingly complex world we've created, it seemed only natural that other fields would have tackled similar problems we could learn from and appropriate. As it turns out, loads of other fields such as industrial design and architecture[40] have developed smart modular systems for manufacturing immensely complex objects like airplanes, ships, and skyscrapers.

But my original explorations kept creeping back to the natural world, which triggered memories of sitting at a rickety desk in my high school's chemistry lab.

Taking cues from chemistry

My high school chemistry class was taught by a no-nonsense Vietnam vet with an extraordinarily impressive mustache. Mr. Rae's class had a reputation for being one of the hardest classes in school, largely because of an assignment that required students to balance hundreds upon hundreds of chemical equations contained in a massive worksheet.

If you're like me, you may need a bit of a refresher to recall what a chemical equation looks like, so here you go:

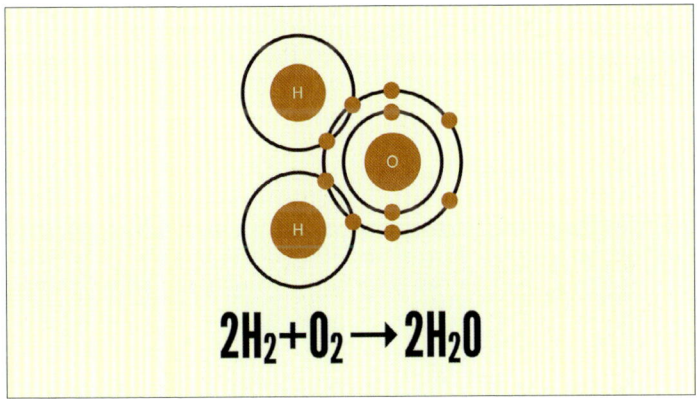

An example of a chemical equation showing hydrogen and oxygen atoms combining together to form a water molecule.

40   http://us5.campaign-archive1.com/?u=7e093c5cf4&id=ead8a72012&e=ecb25a3f93

Chemical reactions are represented by chemical equations, which often show how atomic elements combine together to form molecules. In the example above, we see how hydrogen and oxygen combine together to form water molecules.

In the natural world, **atomic elements combine together to form molecules. These molecules can combine further to form relatively complex organisms.** To expound a bit further:

- **Atoms** are the basic building blocks of all matter. Each chemical element has distinct properties, and they can't be broken down further without losing their meaning. (Yes, it's true atoms are composed of even smaller bits like protons, electrons, and neutrons, but atoms are the smallest *functional* unit.)

- **Molecules** are groups of two or more atoms held together by chemical bonds. These combinations of atoms take on their own unique properties, and become more tangible and operational than atoms.

- **Organisms** are assemblies of molecules functioning together as a unit. These relatively complex structures can range from single-celled organisms all the way up to incredibly sophisticated organisms like human beings.

Of course, I'm simplifying the incredibly rich composition of the universe, but the basic gist remains: atoms combine together to form molecules, which further combine to form organisms. This atomic theory means that all matter in the known universe can be broken down into a finite set of atomic elements:

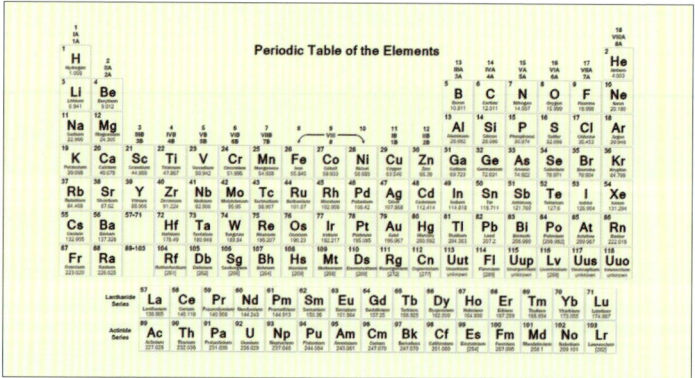

The periodic table of chemical elements.

Apparently Mr. Rae's strategy of having students mind-numbingly balance tons of chemical equations worked, because I'm coming back to it all these years later for inspiration on how to approach interface design.

The atomic design methodology

By now you may be wondering why we're talking about atomic theory, and maybe you're even a bit angry at me for forcing you to relive memories of high school chemistry class. But this is going somewhere, I promise.

We discussed earlier how all matter in the universe can be broken down into a finite set of atomic elements. As it happens, our interfaces can be broken down into a similar finite set of elements. Josh Duck's Periodic Table of HTML Elements[41] beautifully articulates how all of our websites, apps, intranets, hoobadyboops, and whatevers are all composed of the same HTML elements.

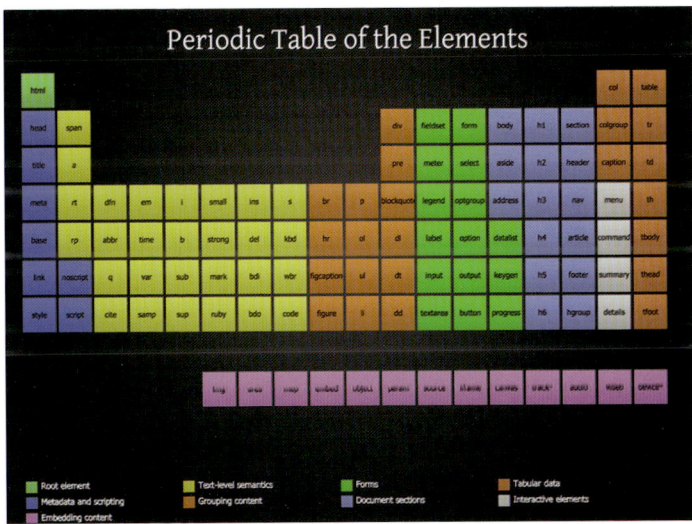

The periodic table of HTML elements by Josh Duck.

41 http://smm.zoomquiet.io/data/20110511083224/index.html

Because we're starting with a similar finite set of building blocks, we can apply that same process that happens in the natural world to design and develop our user interfaces.

Enter atomic design.

Atomic design is a methodology composed of five distinct stages working together to create interface design systems in a more deliberate and hierarchical manner. The five stages of atomic design are:

1. Atoms
2. Molecules
3. Organisms
4. Templates
5. Pages

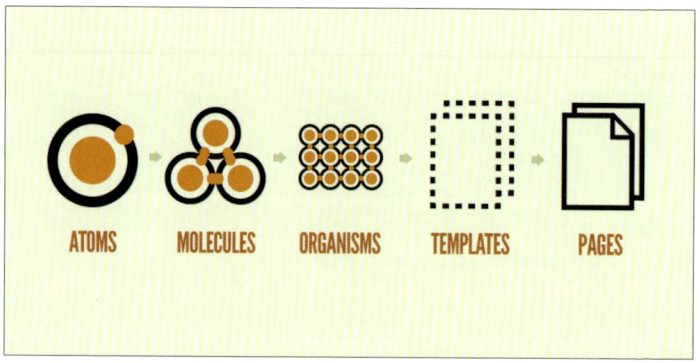

Atomic design is atoms, molecules, organisms, templates, and pages concurrently working together to create effective interface design systems.

Atomic design is not a linear process, but rather a mental model to help us think of our user interfaces as both a cohesive whole and a collection of parts *at the same time*. Each of the five stages plays a key role in the hierarchy of our interface design systems. Let's dive into each stage in a bit more detail.

Atoms

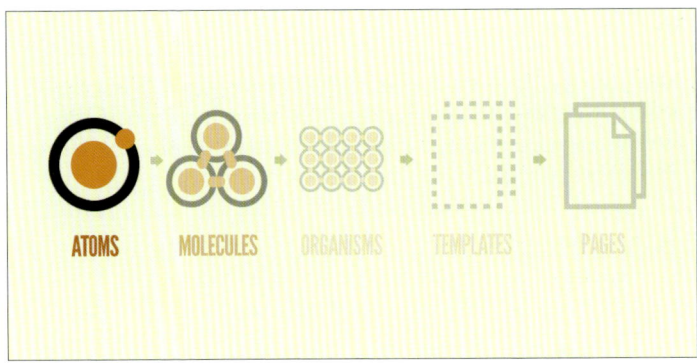

If atoms are the basic building blocks of matter, then the **atoms of our interfaces serve as the foundational building blocks that comprise all our user interfaces.** These atoms include basic HTML elements[42] like form labels, inputs, buttons, and others that can't be broken down any further without ceasing to be functional.

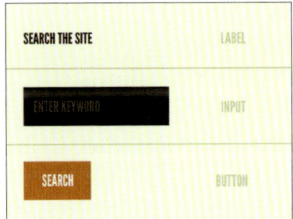

Atoms include basic HTML tags like inputs, labels, and buttons.

Each atom in the natural world has its own unique properties. A hydrogen atom contains one electron, while a helium atom contains two. These intrinsic chemical properties have profound effects on their application (for example, the Hindenburg explosion was so catastrophic because the airship was filled with extremely flammable hydrogen gas versus inert helium gas). In the same

42 https://developer.mozilla.org/en-US/docs/Web/HTML/Element

manner, each interface atom has its own unique properties, such as the dimensions of a hero image, or the font size of a primary heading. These innate properties influence how each atom should be applied to the broader user interface system.

In the context of a pattern library, atoms demonstrate all your base styles at a glance, which can be a helpful reference to keep coming back to as you develop and maintain your design system. But like atoms in the natural world, interface atoms don't exist in a vacuum and only really come to life with application.

Molecules

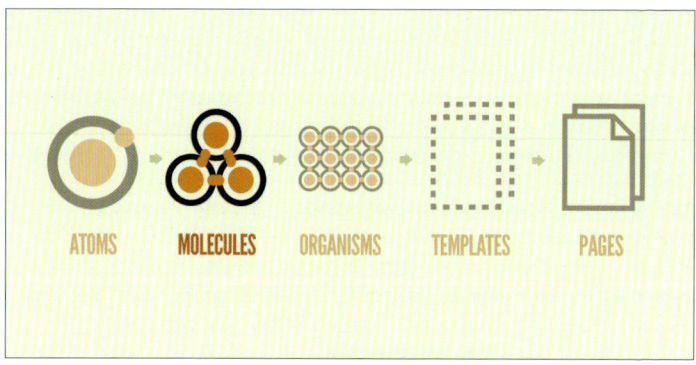

In chemistry, molecules are groups of atoms bonded together that take on distinct new properties. For instance, water molecules and hydrogen peroxide molecules have their own unique properties and behave quite differently, even though they're made up of the same atomic elements (hydrogen and oxygen).

In interfaces, **molecules are relatively simple groups of UI elements functioning together as a unit.** For example, a form label, search input, and button can join together to create a search form molecule.

A search form molecule is composed of a label atom, input atom, and button atom.

When combined, these abstract atoms suddenly have purpose. The label atom now defines the input atom. Clicking the button atom now submits the form. The result is a simple, portable, reusable component that can be dropped in anywhere search functionality is needed.

Now, assembling elements into simple functioning groups is something we've always done to construct user interfaces. But dedicating a stage in the atomic design methodology to these relatively simple components affords us a few key insights.

Creating simple components helps UI designers and developers adhere to the single responsibility principle, an age-old computer science precept that encourages a "do one thing and do it well" mentality. Burdening a single pattern with too much complexity makes software unwieldy. Therefore, creating simple UI molecules makes testing easier, encourages reusability, and promotes consistency throughout the interface.

Now we have simple, functional, reusable components that we can put into a broader context. Enter organisms!

Organisms

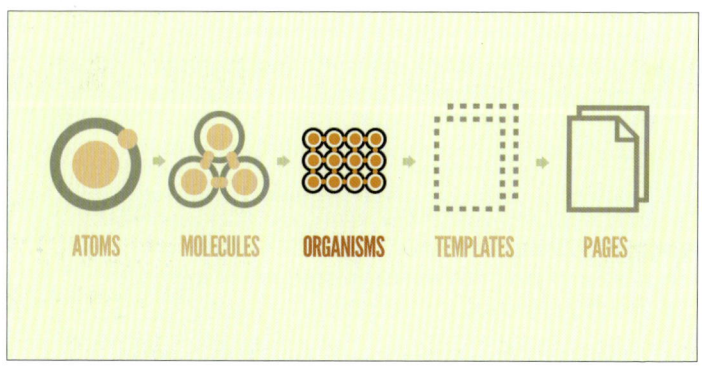

Organisms are relatively complex UI components composed of groups of molecules and/or atoms and/or other organisms. These organisms form distinct sections of an interface.

Let's revisit our search form molecule. A search form can often be found in the header of many web experiences, so let's put that search form molecule into the context of a header organism.

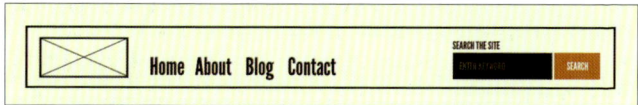

This header organism is composed of a search form molecule, logo atom, and primary navigation molecule.

The header forms a standalone section of an interface, even though it contains several smaller pieces of interface with their own unique properties and functionality.

Organisms can consist of similar or different molecule types. A header organism might consist of dissimilar elements such as a logo image, primary navigation list, and search form. We see these types of organisms on almost every website we visit.

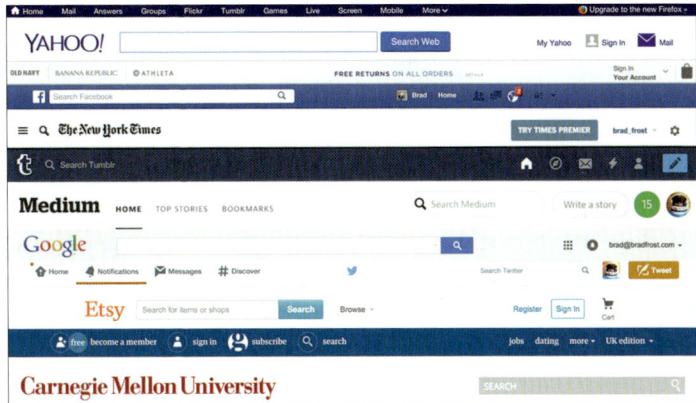

Organisms like website headers consist of smaller molecules like primary navigation, search forms, utility navigation, and logos.

While some organisms might consist of different types of molecules, other organisms might consist of the same molecule repeated over and over again. For instance, visit a category page of almost any e-commerce website and you'll see a listing of products displayed in some form of grid.

Building up from molecules to more elaborate organisms provides designers and developers with an important sense of context. Organisms demonstrate those smaller, simpler components in action and serve as distinct patterns that can be used again and again. The product grid organism can be employed anywhere a group of products needs to be displayed, from category listings to search results to related products.

Now that we have organisms defined in our design system, we can break our chemistry analogy and apply all these components to something that resembles a web page!

A product grid organism on Gap's e-commerce website consists of the same product item molecule repeated again and again.

Templates

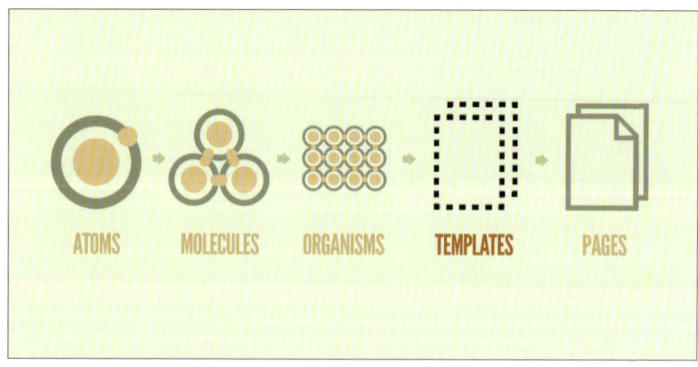

Now, friends, it's time to say goodbye to our chemistry analogy. The language of atoms, molecules, and organisms carries with it a helpful hierarchy for us to deliberately construct the components of our design systems. But ultimately we must step into language that is more appropriate for our final output and makes more sense to our clients, bosses, and colleagues. Trying to carry the chemistry analogy too far might confuse your stakeholders and cause them to think you're a bit crazy. Trust me.

Templates are page-level objects that place components into a layout and articulate the design's underlying content structure. To build on our previous example, we can take the header organism and apply it to a homepage template.

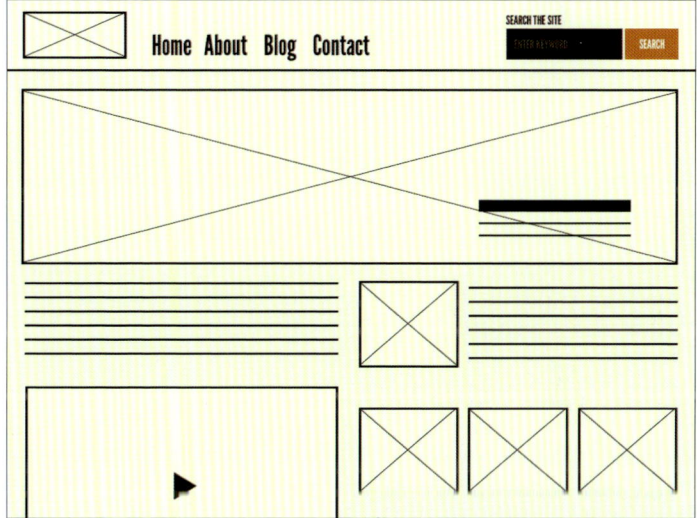

The homepage template consists of organisms and molecules applied to a layout.

This homepage template displays all the necessary page components functioning together, which **provides context for these relatively abstract molecules and organisms**. When crafting an effective design system, it's critical to demonstrate how components look and function together in the context of a layout

CHAPTER 2 / ATOMIC DESIGN METHODOLOGY 51

to prove the parts add up to a well-functioning whole. We'll discuss this more in a bit.

Another important characteristic of templates is that they **focus on the page's underlying content structure** rather than the page's final content. Design systems must account for the dynamic nature of content, so it's very helpful to articulate important properties of components like image sizes and character lengths for headings and text passages.

Mark Boulton discusses the importance of defining the underlying content structure of a page:

> *You can create good experiences without knowing the content. What you can't do is create good experiences without knowing your content structure. What is your content made from, not what your content is.*
>
> – Mark Boulton[43]

By defining a page's skeleton we're able to create a system that can account for a variety of dynamic content, all while providing needed guardrails for the types of content that populate certain design patterns. For example, the homepage template for Time Inc. shows a few key components in action while also demonstrating content structure regarding image sizes and character lengths:

---

43 http://www.markboulton.co.uk/journal/structure-first-content-always

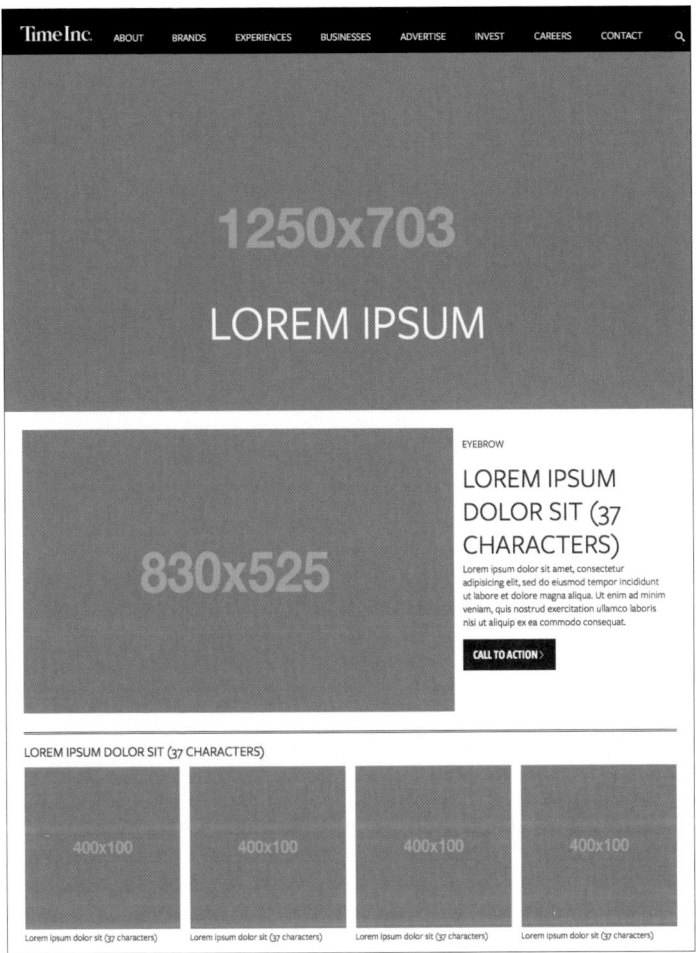

Time Inc.'s homepage template demonstrates the content's underlying structure.

Now that we've established our pages' skeletal system, let's put some meat on them bones!

Pages

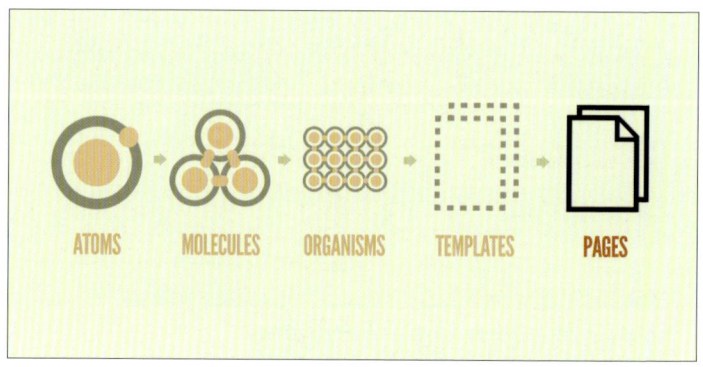

Pages are specific instances of templates that show what a UI looks like with real representative content in place. Building on our previous example, we can take the homepage template and pour representative text, images, and media into the template to show real content in action.

The page stage is the most concrete stage of atomic design, and it's important for some rather obvious reasons. After all, this is what users will see and interact with when they visit your experience. This is what your stakeholders will sign off. And this is where you see all those components coming together to form a beautiful and functional user interface. Exciting!

The page stage replaces placeholder content with real representative content to bring the design system to life.

In addition to demonstrating the final interface as your users will see it, **pages are essential for testing the effectiveness of the underlying design system**. It is at the page stage that we're able to take a look at how all those patterns hold up when real content is applied to the design system. Does everything look great and function as it should? If the answer is no, then we can loop back and modify our molecules, organisms, and templates to better address our content's needs.

When we pour real representative content into Time Inc.'s homepage template, we're able to see how all those underlying design patterns hold up.

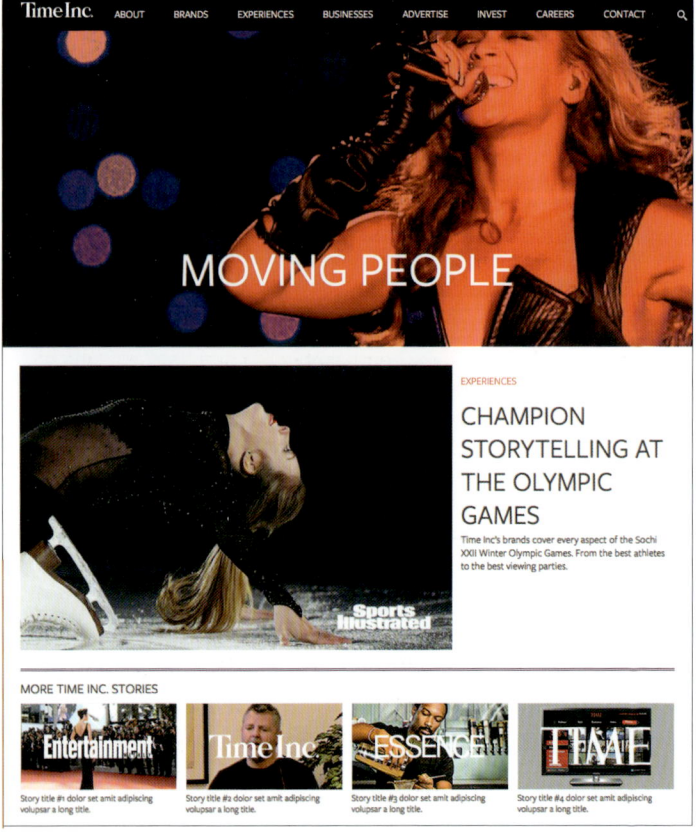

At the page stage, we're able to see what Time Inc.'s homepage looks like with real representative content in place. With actual content in place, we're able to see if the UI components making up the page properly serve the content being poured into them.

We must create systems that establish reusable design patterns and also accurately reflect the reality of the content we're putting inside of those patterns.

Pages also provide a place to articulate variations in templates, which is crucial for establishing robust and reliant design systems. Here are just a few examples of template variations:

- A user has one item in their shopping cart and another user has ten items in their cart.
- A web app's dashboard typically shows recent activity, but that section is suppressed for first-time users.
- One article headline might be 40 characters long, while another article headline might be 340 characters long.
- Users with administrative privileges might see additional buttons and options on their dashboard compared to users who aren't admins.

In all of these examples, the underlying templates are the same, but the user interfaces change to reflect the dynamic nature of the content. These variations directly influence how the underlying molecules, organisms, and templates are constructed. Therefore, creating pages that account for these variations helps us create more resilient design systems.

So that's atomic design! These five distinct stages concurrently work together to produce effective user interface design systems. To sum up atomic design in a nutshell:

- **Atoms** are UI elements that can't be broken down any further and serve as the elemental building blocks of an interface.
- **Molecules** are collections of atoms that form relatively simple UI components.
- **Organisms** are relatively complex components that form discrete sections of an interface.
- **Templates** place components within a layout and demonstrate the design's underlying content structure.
- **Pages** apply real content to templates and articulate variations to demonstrate the final UI and test the resilience of the design system.

Advantages of atomic design

So why go through all this rigamarole? What's atomic design good for? These are valid questions, considering we've been building user interfaces for a long time now without having an explicit five-stage methodology in place. But atomic design provides us with a few key insights that help us create more effective, deliberate UI design systems.

The part and the whole

One of the biggest advantages atomic design provides is the ability to quickly shift between abstract and concrete. We can simultaneously see our interfaces broken down to their atomic elements and also see how those elements combine together to form our final experiences.

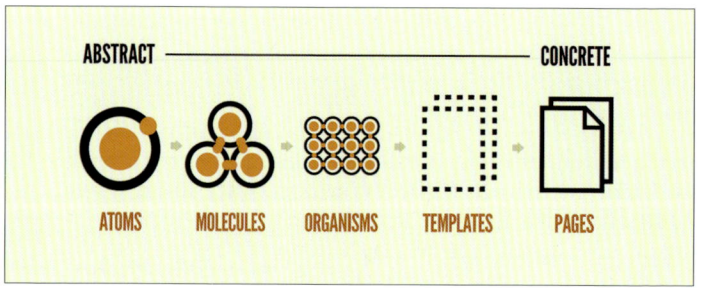

Atomic design allows designers to traverse between abstract and concrete.

In his book *The Shape of Design*, Frank Chimero beautifully articulates the power this traversal provides:

> The painter, when at a distance from the easel, can assess and analyze the whole of the work from this vantage. He scrutinizes and listens, chooses the next stroke to make, then approaches the canvas to do it. Then, he steps back again to see what he's done in relation to the whole. It is a dance of switching contexts, a pitter-patter pacing across the studio floor that produces a tight feedback loop between mark-making and mark-assessing.
>
> – Frank Chimero[44]

44 http://read.shapeofdesignbook.com/chapter01.html

Atomic design lets us dance between contexts like the painter Frank so eloquently describes. The atoms, molecules, and organisms that comprise our interfaces do not live in a vacuum. And our interfaces' templates and pages are indeed composed of smaller parts. The parts of our designs influence the whole, and the whole influences the parts. The two are intertwined, and atomic design embraces this fact.

When designers and developers are crafting a particular component, we are like the painter at the canvas creating detailed strokes. When we are viewing those components in the context of a layout with real representative content in place, we are like the painter several feet back from the canvas assessing how their detailed strokes affect the whole composition. It's necessary to zero in on one particular component to ensure it is functional, usable, and beautiful. But it's also necessary to ensure that component is functional, usable, and beautiful *in the context of the final UI*.

Atomic design provides us a structure to navigate between the parts and the whole of our UIs, which is why it's crucial to reiterate that **atomic design is not a linear process**. It would be foolish to design buttons and other elements in isolation, then cross our fingers and hope everything comes together to form a cohesive whole. So don't interpret the five stages of atomic design as "Step 1: atoms; Step 2: molecules; Step 3: organisms; Step 4: templates; Step 5: pages." Instead, **think of the stages of atomic design as a mental model that allows us to concurrently create final UIs and their underlying design systems.**

Clean separation between structure and content

Discussing *design* and *content* is a bit like discussing the *chicken* and the *egg*. Mark Boulton explains:

> Content needs to be structured and structuring alters your content, designing alters content. It's not 'content then design', or 'content or design'. It's 'content and design'.
>
> - Mark Boulton[45]

---

45 http://www.markboulton.co.uk/journal/structure-first-content-always

A well-crafted design system caters to the content that lives inside it, and well-crafted content is aware of how it's presented in the context of a UI. The interface patterns we establish must accurately reflect the nature of the text, images, and other content that live inside them. Similarly, our content should be aware of the manner in which it will be presented. The close relationship between content and design requires us to consider both as we construct our UIs.

Atomic design gives us a language for discussing the structure of our UI patterns and also the content that goes inside those patterns. While there is a clean separation between the content structure skeleton (templates) and the final content (pages), atomic design recognizes the two very much influence each other. For instance, take the following example:

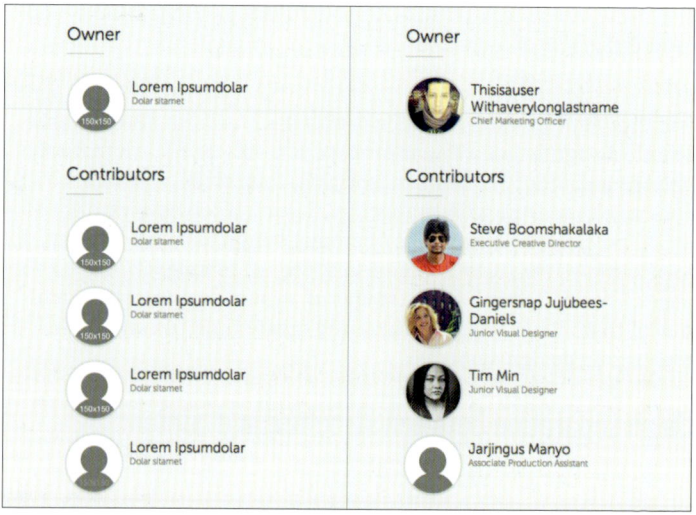

On the left we see the UI's content skeleton, which consists of the same *person block molecule* repeated again and again. On the right we see what happens when we populate each instance of the person block molecule with representative content. Visualizing the content skeleton and the representative final content allows us to create patterns that accurately reflect the content that lives inside them. If a person's name were to wrap onto five lines within the pattern, we would need to address that broken behavior at a more atomic level.

The content we pour into our UIs at the page stage will influence the characteristics and parameters of the underlying design patterns.

What's in a name?

Throughout this book I've mentioned that modular design and development is nothing new. So why are we introducing terms like *atoms, molecules,* and *organisms* when we can just stick with established terms like *modules, components, elements, sections,* and *regions*?

For as long as I've been talking about atomic design, I've had people proffer alternate names for the stages of the methodology. Person One would suggest, "Why not just name them elements, modules, and components?" while Person Two would suggest, "Why not just name them base, components, and modules?" The issue with terms like components and modules is that a sense of hierarchy can't be deduced from the names alone. **Atoms, molecules, and organisms imply a hierarchy** that anyone with a basic knowledge of chemistry can hopefully wrap their head around.

That being said, naming things is hard and imperfect. The names I've chosen for the stages of atomic design have worked really well for me and the teams I've worked with as we create UI design systems. But maybe they don't work for you and your organization. That's more than OK. Here's one perspective from the design team at GE:

> As we showed our initial design system concepts that used the Atomic Design taxonomy to our colleagues, we were met with the some confused looks. […] The evidence was clear, for this to be successful within our organization we had to make the taxonomy more approachable.
>
> – Jeff Crossman, GE Design[46]

The taxonomy the team landed on were "Principles", "Basics", "Components", "Templates", "Features", and "Applications". Do these labels make sense to you? It doesn't matter. By establishing a taxonomy that made sense for their organization, everyone was able to get on board with atomic design principles and do effective work together.

[46] https://medium.com/ge-design/ges-predix-design-system-8236d47b0891#.uo68yjo9g

"Atomic design" as a buzzword encapsulates the concepts of modular design and development, which becomes a useful shorthand for convincing stakeholders and talking with colleagues. But **atomic design is not rigid dogma**. Ultimately, whatever taxonomy you choose to work with should help you and your organization communicate more effectively in order to craft an amazing UI design system.

Atomic design is for user interfaces

Atomic design is a concept born of the web. After all, your lowly author is a web designer, which is mainly the reason this book primarily focuses on web-based interfaces. But it's important to understand that **atomic design applies to all user interfaces, not just web-based ones.**

You can apply the atomic design methodology to the user interface of any software: Microsoft Word, Keynote, Photoshop, your bank's ATM, whatever. To demonstrate, let's apply atomic design to the native mobile app Instagram.

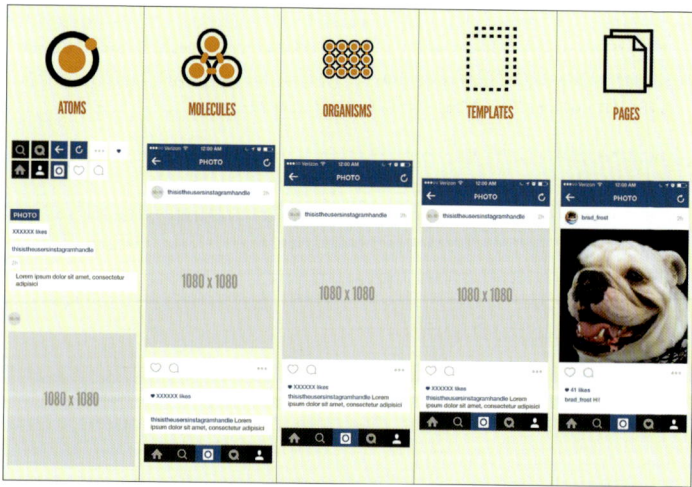

Atomic design applied to the native mobile app Instagram.

Let's walk through this atomized Instagram interface:

- **Atoms:** This screen of Instagram's UI consists of a handful of icons, some text-level elements, and two image types: the primary image and the user's avatar image.

- **Molecules:** Several icons form simple utilitarian components like the bottom navigation bar and the photo actions bar where users can like or comment on a photo. Also, simple combinations of text and/or images form relatively simple components.

- **Organisms:** Here we can see the photo organism take shape, which consists of the user's information, time stamp, the photo itself, actions around that photo, and information about the photo including like count and caption. This organism becomes the cornerstone of the entire Instagram experience as it is stacked repeatedly in a never-ending stream of user-generated photos.

- **Templates:** We get to see our components come together in the context of a layout. Also, it's here where we see the exposed content skeleton of the Instagram experience, highlighting dynamic content such as the user's handle, avatar, photo, like count, and caption.

- **Pages:** And finally we see the final product, complete with real content poured into it, which helps ensure the underlying design system comes together to form a beautiful and functional UI.

I show this non-web example because atomic design tends to get misinterpreted as an approach to web-specific technologies like CSS and JavaScript. Let me be clear about this: **atomic design has nothing to do with web-specific subjects like CSS or JavaScript architecture.** In chapter 1 we discussed the trend toward modularity in all aspects of design and development, which includes CSS and JavaScript. These are fantastic trends in CSS and JavaScript, but atomic design deals with crafting user interface design systems irrespective of the technology used to create them.

Atomic design in theory and in practice

This chapter introduced the atomic design methodology and demonstrated how atoms, molecules, organisms, templates, and pages all work together to craft thoughtful, deliberate interface design systems. Atomic design allows us to see our UIs broken down to their atomic elements, and also allows us to simultaneously step through how those elements join together to form our final UIs. We learned about the tight bond between content and design, and how atomic design allows us to craft design systems that are tailored to the content that lives inside them. And finally we learned how the language of atomic design gives us a helpful shorthand for discussing modularity with our colleagues, and provides a much needed sense of hierarchy in our design systems.

Atomic design is a helpful design and development methodology, but essentially it's merely a mental model for constructing a UI. By now you may be wondering *how* you make atomic design happen. Well, fear not, dear reader, because the rest of the book focuses on tools and processes to make your atomic design dreams come true.

Chapter 3
Tools of the Trade

Pattern Lab and the qualities of effective style guides

In the previous chapter, I introduced the atomic design methodology for constructing user interfaces. I hope you'll find atomic design to be a helpful mental model for constructing UI design systems, but now it's time to climb down from the ivory tower and actually put atomic design into practice *in the real world*.

The cornerstone of pattern-based design and development is the pattern library, which serves as a centralized hub of all the UI components that comprise your user interface. As we discussed in chapter 1, the benefits of pattern libraries are many:

- They **promote consistency and cohesion** across the entire experience.
- They **speed up your team's workflow**, saving time and money.
- They **establish a more collaborative workflow** between all disciplines involved in a project.
- They **establish a shared vocabulary** between everyone in an organization, including outside vendors.
- They **provide helpful documentation** to help educate stakeholders, colleagues, and even third parties.
- They **make cross-browser/device, performance, and accessibility testing easier**.
- They **serve as a future-friendly foundation** for teams to modify, extend, and improve on over time.

That all sounds wonderful, right? I can almost hear you saying, "I need this whole pattern library thing in my life." But how do we make pattern libraries happen? Well, you've come to the right place, friend, because the rest of this book is dedicated to exactly that. This chapter will introduce helpful tools for creating pattern libraries, the next chapter will discuss how to make patterns a cornerstone of your design and development workflow, and the fifth chapter will cover how to make your design system stand the test of time.

This chapter will talk about the qualities of effective pattern libraries through the lens of a tool called Pattern Lab[47], an open source project maintained by web developers Dave Olsen[48],

47 http://patternlab.io/
48 http://dmolsen.com/

[Brian Muenzenmeyer](http://www.brianmuenzenmeyer.com/)[49], and me to execute atomic design systems. While I'll excitedly discuss Pattern Lab and its various features, I want to stress that the point of this chapter is to cover the characteristics of well-constructed pattern libraries, not sell any one specific tool to you. Hell, Pattern Lab isn't even for sale! No single tool will be a perfect fit for every setup and scenario, but be sure to keep the following principles in mind when deciding what tools to use to create your pattern libraries.

Just what exactly is Pattern Lab?

Before we dive into the nuts and bolts of how Pattern Lab works, it's important to take time to explain what the tool is and isn't.

Pattern Lab is…

- a static site generator tool for building atomic design systems.
- a pattern documentation and annotation tool.
- a pattern starter kit.

Pattern Lab isn't…

- a UI framework like Bootstrap or Foundation.
- language-, library-, or style-dependent.
- a replacement for a content management system.

Let's walk through these points, starting with the term *static site generator*. Static site generator tools take in some source code and assets, compile them, and spit out plain ol' HTML, CSS, and JavaScript at the other end. **Pattern Lab takes source code – namely patterns – and compiles those patterns into a functional front-end UI inside a pattern library shell**.

So what does Pattern Lab look like out of the box? Drumroll, please.

---

49 http://www.brianmuenzenmeyer.com/

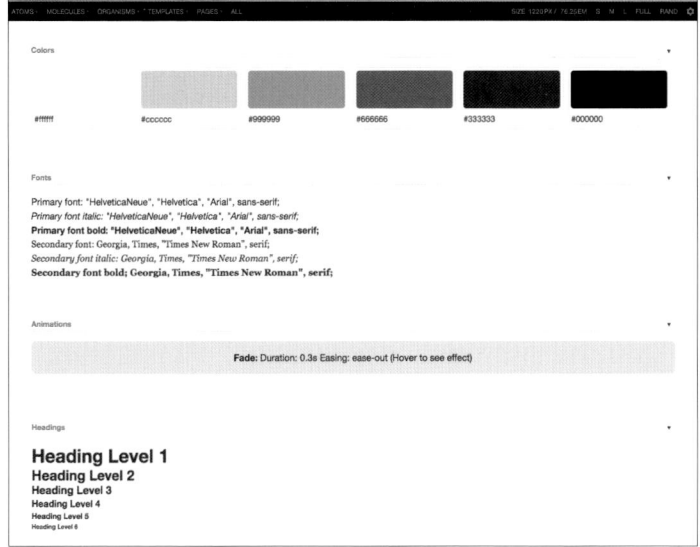

One default Pattern Lab dashboard. What it lacks in good looks, it makes up for in utility.

Not a terribly inspiring design, eh? Believe it or not, this minimal (one might even say *lack* of) design is deliberate. To avoid incorrect classification as a UI framework like Bootstrap, the design is deliberately stripped down so no one would mistakenly take Pattern Lab's demo UI for suggested styles. Pattern Lab doesn't give you any answers as to how to design or architect your front-end code — *you have to do all that work yourself.* The look and feel, naming conventions, syntax, structure, libraries, and scripts you choose to use to create your UI are entirely up to you and your team. Heck, you can even use UI frameworks like Bootstrap *within* Pattern Lab. Pattern Lab is just there to help stitch everything together.

As a technical aside, Pattern Lab uses either PHP or Node.js as the engine that stitches patterns together and generates the pattern library. However, you don't need to be a PHP wizard or Node.js guru to use Pattern Lab any more than you have to know how to build an internal combustion engine to drive a car. Moreover, your final website doesn't have to be built with PHP or Node.js to use the tool, as Pattern Lab's output is backend-agnostic HTML, CSS, and JavaScript. So like any technology decision, choose a pattern library tool that fits with your organization's infrastructure and jives with how your team works together.

If that all sounded like gibberish to you, don't worry. This chapter focuses on the overarching features of Pattern Lab and principles of effective pattern libraries rather than going too far down the technical rabbit hole. If interested, you can check out Pattern Lab's documentation[50] to dive into the nitty-gritty.

Building atomic design systems with Pattern Lab

To understand the core concept behind Pattern Lab, you need to understand Russian nesting dolls.

Russian nesting dolls. Image credit: S. Faric on Flickr[51]

Matryoshka dolls (also known as Russian nesting dolls) are beautifully carved hollow wooden dolls of increasing size that are placed inside one another. Patterns in Pattern Lab operate in a similar manner: the smallest patterns (atoms) are included inside bigger patterns (molecules), which are included in even bigger patterns (organisms), which are in turn included in even bigger patterns (templates).

50 http://patternlab.io/docs/
51 https://www.flickr.com/photos/tromal/6901848291/

Constructing UIs in this manner keeps things DRY[52], which is a long-standing computer science principle that stands for "don't repeat yourself." What this means is that you can make a change to a pattern, and anywhere that pattern is employed will magically update with those changes. This Russian nesting doll approach considerably speeds up your workflow, and certainly beats the pants off sifting through hundreds of Photoshop documents for every instance of a pattern in order to make a simple change.

To make this happen, Pattern Lab uses the *include* feature of Mustache, a logicless templating language. Here's what a Mustache include looks like:

```
{{> atom-thumbnail }}
```

This is Mustache code, in case the double curly braces ({{}}) that look like little mustaches didn't give it away. The greater-than symbol (>) is Mustache's way of telling Pattern Lab to include an atom pattern called "thumbnail". Pattern Lab will go searching through its folders of patterns to find an atom named "thumbnail".

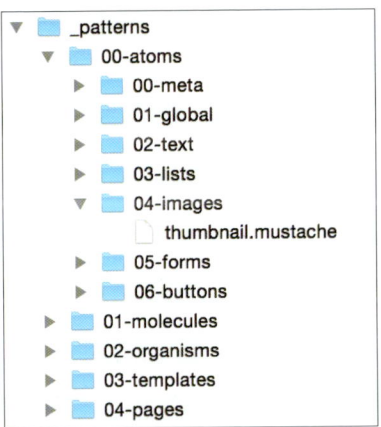

This is what Pattern Lab's patterns folder structure can look like. You can name and categorize these folders however you'd like, including changing the labels "atoms", "molecules", and "organisms", "templates", and "pages". The most important consideration is to establish naming and categorization that is most effective for your team.

52 https://en.wikipedia.org/wiki/Don't_repeat_yourself

Now that we know what an include looks like, let's put it into practice and take a look at a few patterns from a website I helped make for Time Inc. Here's one reusable pattern we designed:

For Time Inc.'s website, we created a basic block molecule consisting of a thumbnail image, headline, and excerpt.

This pattern should look fairly familiar. A thumbnail image, headline, and excerpt working together as a single unit is a common pattern found on countless websites. Let's take a peek behind the curtain to see how this pattern is constructed.

```
<div class="block-post">
    <a href="{{ url }}">
        {{> atoms-thumb }}
        <h3>{{ headline }}</h3>
        <p>{{ excerpt }}</p>
    </a>
</div>
```

You can see we have: HTML markup consisting of a wrapper `div` with a class name of `block-post`; a link; a Mustache include for the thumbnail image; an `<h3>` element for the headline; and a `<p>` tag for the excerpt. You'll notice there's more Mustache code for `url`, `headline`, and `excerpt`, which we'll use later to dynamically swap in actual content. More on that in a bit.

Now that our pattern markup is established, we can include that chunk of code in bigger patterns using the same include method:

```
{{> molecules-block-post }}
```

Now let's move up to more complex organisms like the website's header, which looks a little something like this:

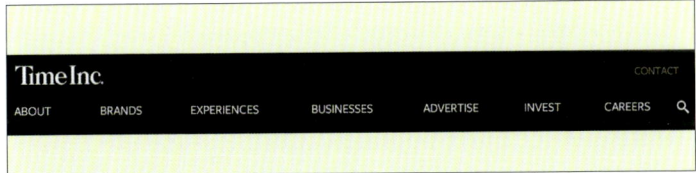

The website header consists of fairly common conventions like a logo atom, primary navigation molecule, and a search form molecule.

When we crack open the hood to look at the header's markup in Pattern Lab, we see the following:

```
<header role="banner">
    {{> atoms-logo }}
    {{> molecules-primary-nav }}
    {{> molecules-search }}
</header>
```

What's going on here? Well, we have a basic `<header>` element, and inside that element we're including the logo image atom, the primary navigation molecule, and the search form molecule.

And now we can include that relatively complex pattern anywhere we need it.

```
{{> organisms-header }}
```

I hope by now you can see the Russian nesting dolls in action. The smallest atoms are included in bigger molecules, and those molecules get included in even bigger organisms. Now let's take these components and plug them into a layout. Take the homepage template, for instance:

The Time Inc. homepage template consists of a few repeatable patterns: a global header, a hero area, a few sections (containing an image, headline, excerpt, and call to action), an area featuring four items, a factoid area, and a global footer.

Take a quick stroll through the homepage template and you'll see some pretty standard patterns: a site header at the top, a site footer at the bottom, and a full-screen hero area. You'll also see a few other patterns repeating themselves throughout the template.

So how does this look in code? As you might expect, it involves more includes!

```
{{> organisms-header }}
<main role="main">
    {{# hero }}
    {{> molecules-hero }}
    {{/ hero }}
    <section>
        {{# experience-block }}
        {{> molecules-block-main }}
        {{/ experience-block }}
        {{# experience-feature }}
        {{> organisms-story-feature }}
        {{/ experience-feature }}
    </section>
    <section>
        {{# factoid-advertising }}
        {{> organisms-factoid }}
        {{/ factoid-advertising }}
    </section>
    <section>
        {{# advertising }}
        {{> molecules-block-main }}
        {{/ advertising }}
    </section>
    ...
</main>
{{> organisms-footer }}
```

At this stage in the game the smaller patterns are already constructed, so all the template needs to do is pull them into the context of a page layout and give them unique names.

Taking a closer look at the code, notice that certain patterns like `{{> organisms-header }}` and `{{> organisms-footer }}` are included the same way as the earlier examples. But there are also a few other includes patterns that are supplemented by some additional information, like the following:

```
{{# factoid-advertising }}
{{> organisms-factoid }}
{{/ factoid-advertising }}
```

We're including `organisms-factoid` in the same way as all the other patterns, but we're also naming it `factoid-advertising` by wrapping the include in a Mustache *section*, indicated by the Mustache code containing the # and / symbols. By giving the pattern instance a unique name, we can latch on to it and dynamically replace the content of the pattern. More on that in the next section!

This Russian nesting doll approach to building UIs is simple but tremendously powerful. The structure allows designers and developers to keep patterns DRY, saving time, effort, and money. The approach allows teams to build a final UI while simultaneously creating the underlying UI design system. After all, the final interface is one instantiation of its underlying design system. Teams can also move between abstract and concrete, zeroing in on a particular pattern to fix bugs ("The header's broken!"), while also seeing how changes to small patterns affect the overall page layout.

Working with dynamic data

It's important to articulate the underlying content structure of UI patterns within the context of a pattern library. That's why we've been looking at dimension-displaying grayscale images and placeholder text containing character limits. But while this information is helpful for creative teams, grayscale images and *Lorem ipsum* text are not what users interact with on your actual site. We need a way to replace our dummy content with real representative content to ensure our UI patterns match the reality of the content that lives inside them.

To demonstrate how Pattern Lab dynamically swaps in real content into templates, let's take a look at a side-by-side comparison of Time Inc.'s homepage template and page levels:

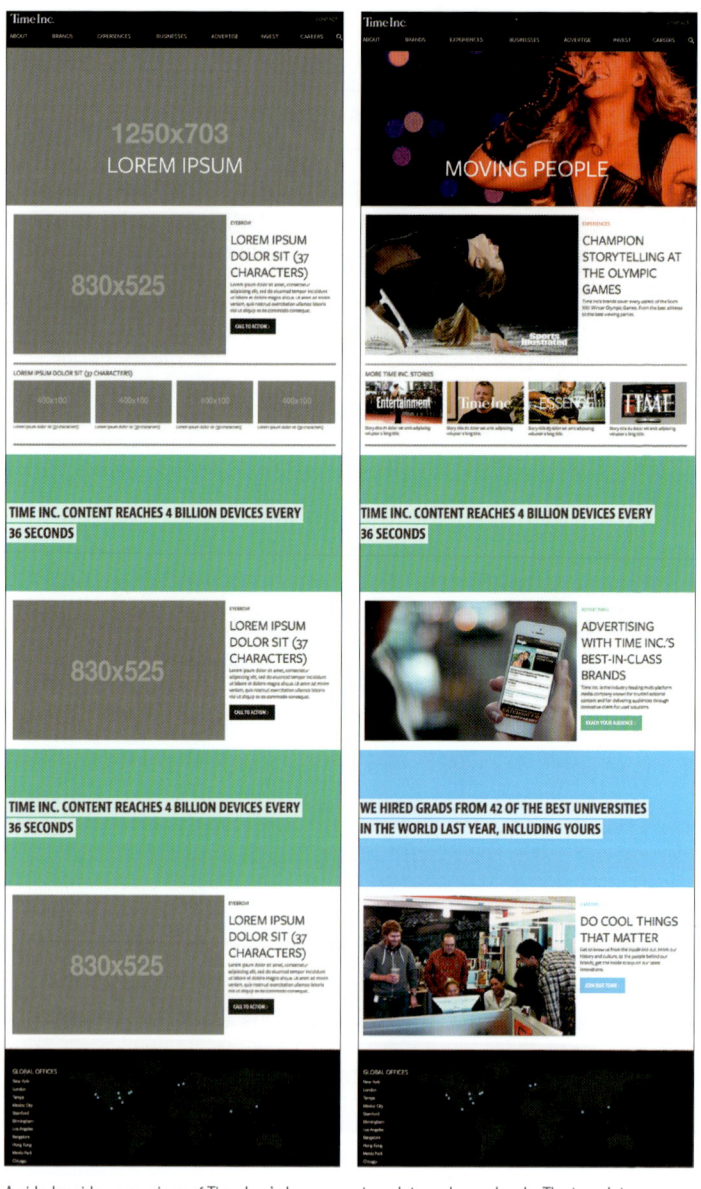

A side-by-side comparison of Time Inc.'s homepage template and page levels. The template articulates the content structure of the design system, while the page shows what the system looks like with real content displayed by it.

On the left we have the template level, which articulates the content structure of the patterns making up the web page. And on the right we have the page level, where we're pouring in real representative content to demonstrate what the final UI might look like and test the effectiveness of the design system.

How do we swap dummy content for real content in Pattern Lab? Pattern Lab uses JSON (as well as YAML, Markdown, and other data formats) to define and swap out the dynamic bits of content in our designs.

The default placeholder data is defined in a file called *data.json* that lives in Pattern Lab's */source* directory. Inside this file we define all the text, image paths, and other dynamic data that will make up our UI. Here's a small sample from Time Inc.'s *data.json* file:

```json
"hero" : {
  "headline": "Lorem Ipsum",
  "img": {
    "src": "/images/sample/fpo_hero.png",
    "alt": "Hero Image"
  }
}
```

For developers, this type of format most likely looks familiar. If you're not a developer, don't freak out! Once you look beyond the curly braces and quotes, you'll see that we're defining a `hero` object (for the full-bleed hero area directly below the header) that has a `headline` value of "Lorem Ipsum", and an `img` with a `src` value of "/images/sample/fpo_hero.png". We're simply defining this object's attributes and providing values for those attributes.

Once those objects are defined, we can then override their attributes at Pattern Lab's page level. This is accomplished by creating a new JSON file that matches the page pattern name (for Time Inc.'s homepage, we'll call it *00-homepage.json*) inside the */pages* directory.

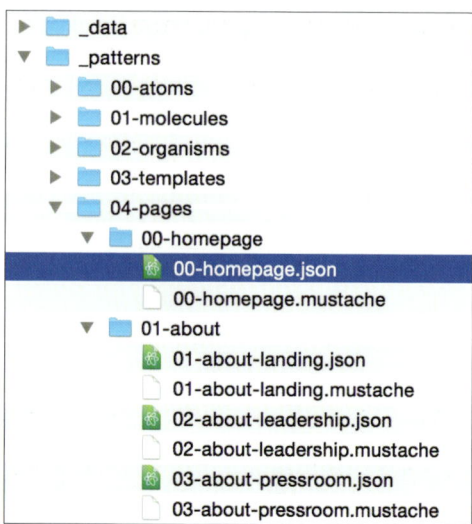

Inside the 'pages' directory we have the homepage pattern as well as a JSON file that matches the name of the pattern. This is where we'll override the default content with page-specific content.

When we open up *00-homepage.json* we can override the placeholder data we established earlier. Here's what that might look like:

```
"hero" : {
  "headline": "Moving People",
  "img": {
    "src": "/images/hero_beyonce.jpg",
    "alt": "Beyonce"
  }
}
```

By overriding the default data, the `hero` headline now reads "Moving People" instead of "Lorem Ipsum." And instead of pointing to a grayscale for-placement-only (FPO) hero image, we're now pointing to a picture of Beyoncé located at "/images/hero_beyonce.jpg".

This process of establishing defaults for dynamic data then replacing them with page-specific content continues for each section of the website. In addition to replacing simple strings like headings, we can also dynamically set variables to `true` or `false`, loop through an array of items, and more. We can even dramatically alter the UI with just a few changes to a JSON file, which we'll talk about next.

Articulating pattern variations with pseudo-patterns

Historically, designers working with static tools have had a tendency to only design best-case scenarios. You know what I'm talking about: the user's name is Sara Smith and always fits neatly on one line; her profile picture looks like it was clipped out of a magazine; her profile is completely filled out; the two columns of her profile content are exactly the same height.

Of course, these best-case scenarios rarely, if ever, occur in the real world.

To create more robust and resilient designs, we need to concurrently account for the best situations and the worst – and everything in between.

What if the user doesn't upload a profile picture? What if the user has 87 items in their shopping cart? What if the product has 14 options? What if the blog post title contains 400 characters? What about a returning user? A first-time user? What if the article doesn't have any comments? What if it has seven layers of nested comments? What if we need to display an urgent message on the dashboard?

Articulating these UI variations in a static design tool is an exercise in tediousness and redundancy, which may explain why they're rarely designed. But if we want to create systems that address all the variables and realities of our content, we must take those 'what if' questions into account.

How do we address all manner of UI variation without exhausting ourselves in the process? Pattern Lab's pseudo-pattern[53] feature

---

53 http://patternlab.io/docs/pattern-pseudo-patterns.html

allows us to articulate sometimes wildly different scenarios with just a few changes to our data.

Let's say we're making an app whose dashboard displays a list of project collaborators. The UI might look something like this:

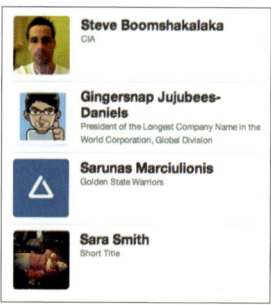

A list of project collaborators in our hypothetical app.

To create the dynamic content inside each of these blocks, we'll define our list of collaborators as an array inside *dashboard.json*:

```
"collaborators": [
  {
    "img": "/images/sample/avatar1.jpg",
    "name": "Steve Boomshakalaka",
    "title": "CIA"
  },
  {
    "img": "/images/sample/avatar2.jpg",
    "name": "Gingersnap Jujubees-Daniels",
    "title": "President of the Longest Company Name in the World Corporation, Global Division"
  },
  {
    "img": "/images/sample/avatar3.jpg",
    "name": "Sarunus Marciulionis",
    "title": "Golden State Warriors"
  },
  {
    "img": "/images/sample/avatar4.jpg",
```

```
        "name"  : "Sara Smith",
        "title" : "Short Title"
    }
]
```

By default, our design assumes the user is a regular user and not an administrator, but what if we wanted to give administrators the ability to manage project collaborators from the dashboard? That UI might look something like this:

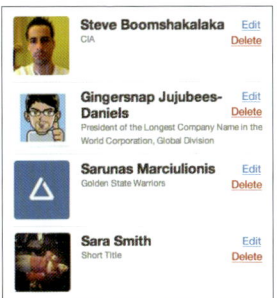

The administrator's dashboard UI introduces extra edit and delete actions.

To show additional admin edit and delete actions on the dashboard in Pattern Lab, we can create a pseudo-pattern, a new file in the /*pages* folder that looks like this:

```
dashboard~admin.json
```

The tilde (~) symbol indicates a pseudo-pattern. *dashboard~admin.json* will inherit all the data contained in *dashboard.json*, but also gives us the opportunity to append or override additional data. That means the list of collaborators we defined earlier in *dashboard.json* is still available, but we can add additional data inside *dashboard~admin.json* like so:

```
"isAdmin" : true
```

We're defining a variable called `isAdmin` and setting it to `true`. We can now use that to conditionally include the additional actions inside the block pattern.

```
<div class="block">
  <img src="{{ img }}" alt="{{ name }}" />
  <h3>{{ name }}</h3>
  <h4>{{ title }}</h4>
  {{# isAdmin }}
  {{> molecules-block-actions }}
  {{/ isAdmin }}
</div>
```

The first few lines are pulling in the img, name, and title we defined in *dashboard.json*. But pay close attention to what's wrapped in the isAdmin Mustache section. What we're saying here is: if isAdmin is set to true, include a molecule pattern called block-actions. The block-actions pattern contains the edit and delete buttons, and will only display if isAdmin is set to true (or anything besides false). In our default *dashboard.json*, isAdmin isn't set, so the extra actions won't display. In *dashboard~admin.json*, we're setting isAdmin to true so the extra actions will display. You can extend this technique to dramatically alter the entire UI (like altering the primary navigation, showing additional panels on the dashboard, adding extra controls, and so on) just by changing a single variable. Powerful stuff, indeed.

Whew. If you've made it this far, congratulations! You now know how to add and manipulate dynamic data in Pattern Lab. Pattern Lab's ability to design with dynamic data provides some crucial benefits:

- **Creates a clear separation between structure and content.** A pattern's structure and its content very much influence each other. However, resilient design systems strive to establish agnostic, flexible patterns that can contain a variety of content. Decoupling pattern structure and data allows us to keep things DRY (which, again, stands for don't repeat yourself) and make changes to content without affecting the pattern structure. Likewise, we're able to make changes to a pattern without having to update every instance of that pattern simply because each instance contains different data. This separation results in huge savings in both time and effort.
- **Establishes an ad hoc CMS.** Establishing default and page-specific data serves as an ad hoc content management system.

As mentioned earlier, static design tools aren't well equipped to handle dynamic data, but it's also overkill to install WordPress, Drupal, or some other CMS just to demonstrate UI variations. Pattern Lab strikes a balance as it allows teams to work with dynamic data but doesn't require setting up any crazy MySQL databases.

- **Serves as a blueprint for back-end developers** responsible for integrating the front-end UI into a real content management system. Back-end developers can look at the UI created in Pattern Lab, understand which bits are static and dynamic, then translate that into the back-end system.
- **Allows writers, content people, designers, and other non-developers to contribute to the living, breathing design system.** As a front-end developer, I can't count the number of times I've had to fix typos, swap in new images, translate copy decks, and make other content-related edits to front-end code. It's death by a million paper cuts, and I'm sure most developers would agree that making minor copy changes isn't an effective use of their time. By separating structure and data, Pattern Lab enables non-developer team members to safely manage the content-related aspects of the design, freeing up developers to focus on building the design system's structure.

We've now covered Pattern Lab's core functionality, but we're not done yet! Next we'll cover a few additional features that should be considered, irrespective of the tool you use to create your pattern library.

Viewport tools for flexible patterns

The multitude of devices now accessing the web has forced designers to re-embrace the intrinsic fluidity of the medium. Thankfully, techniques like responsive web design[54] allow us to create layouts that look and function beautifully on any screen.

It's a no-brainer that we need to establish flexible UI patterns if we want to create responsive designs, but creating fluid patterns has additional advantages. **The more fluid a UI component is, the more resilient and versatile it becomes.** Imagine being able to take a component – let's say a photo gallery slider – and plunk it

54 http://alistapart.com/article/responsive-web-design

anywhere we need it. Sometimes we may need it to be a full-bleed element occupying the entire viewport. Other times we may need to include it in the context of an article. And still other times we may want to include it in the sidebar. The dream is to build our components fluidly and they'll adapt their styles and functionality to fit whatever containers we put them into.

Indeed, this is the promise of container queries[55]. Container queries let elements adapt based on their parent containers rather than the entire viewport, which is how we manipulate elements using media queries at the moment. While still being developed as a native browser capability, container queries will allow us pattern-crazed designers and developers to easily create and deploy fluid UI systems.

Between responsive design, container queries, and good ol'-fashioned common sense, we now understand why it's imperative to create flexible UI patterns. But how do we do that? And how can our pattern library tools help us think and act flexibly?

Many early responsive design testing tools focused on viewing designs on popular mobile device widths, such as 320px (an iPhone 4 in portrait mode), 480px (an iPhone 4 in landscape mode), 768px (an iPad in portrait mode), and so on. But, of course, the web is *much* more diverse than a mobile view, a tablet view, and a desktop view. To help designers better consider the entire resolution spectrum when testing their responsive designs, I created a tool called ish.[56]

The tool is called *ish.* because selecting the small button results in a small-ish viewport. Selecting it again provides a different small-ish viewport. Selecting the medium button gives you a medium-ish viewport. And the large button results in a – wait for it – large-ish viewport. These randomized values help designers and developers better consider the entire resolution spectrum rather than a handful of popular device dimensions.

Ish. is baked into Pattern Lab, which means we can view our UIs and their underlying patterns across the entire resolution spectrum.

While ish. helps designers and developers uncover bugs along the viewport continuum, I've found it to be more helpful as a client and colleague education tool. By building a device-agnostic viewport resizing tool directly into the pattern library, clients and colleagues

55   http://alistapart.com/article/container-queries-once-more-unto-the-breach
56   http://bradfrost.com/demo/ish/

can come to appreciate the fact that their design system should look and function well no matter the viewport size.

Pattern Lab displaying a design in a small-ish viewport.

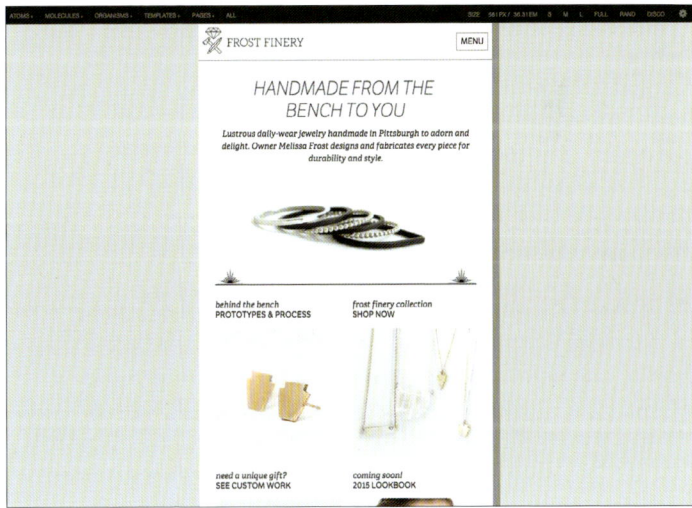

Pattern Lab displaying a design in a medium-ish viewport.

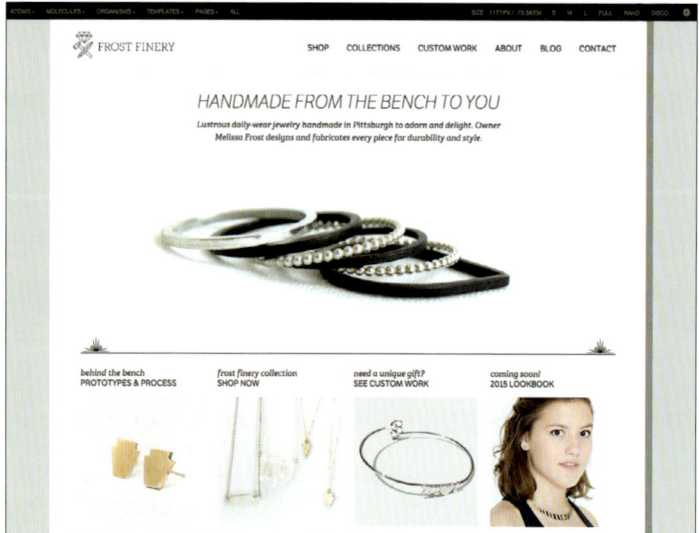

Pattern Lab displaying a design in a large-ish viewport.

A look under the hood with code view

A common pattern library feature is the ability to peek under the hood and view the underlying code that makes up a particular component. Exposing a UI pattern's code speeds up development time (I love copying and pasting as much as the next coder) and helps team leaders enforce code syntax and style conventions. This becomes especially rewarding when a ton of developers are working on an organization's codebase.

The types of code to be highlighted in a pattern library naturally vary from organization to organization, in order to meet the requirements of the vast array of environments, technologies, and conventions used. Most pattern libraries out there in the wild[57] demonstrate a pattern's underlying HTML, while others also include pattern-specific CSS and JavaScript. Salesforce's Lightning design system, for example, shows a pattern's HTML as well as all the (S)CSS pertaining to that pattern.

57 http://styleguides.io/examples.html

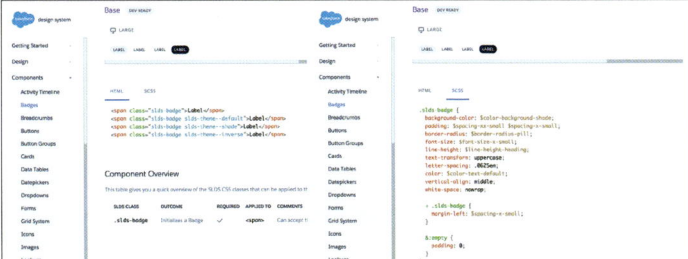

Salesforce's Lightning design system showcases the UI components' HTML and SCSS code.

Including front-end code leads to authors writing it more consistently, but that doesn't guarantee perfection. There's still room for developers to go rogue and write sloppy, incongruent code – which is why some organizations have gone above and beyond to establish incredibly sophisticated design systems. Companies like Lonely Planet have achieved the holy grail of pattern libraries, which is to say **their pattern library and production environment are perfectly in sync**. We'll discuss the holy grail in more detail in chapter 5, but it's worth bringing up in this section to demonstrate how that affects the code exposed in the context of a pattern library. Rather than offering HTML and CSS, Lonely Planet's Rizzo style guide[58] surfaces the include code for teams to pull in the appropriate UI component.

Lonely Planet's Rizzo design system pattern library showcases the template usage.

58 http://rizzo.lonelyplanet.com/

This setup allows the core development team to maintain a single source of truth for all patterns' front-end code. For developers to get up and running, the pattern library needs only provide the code to include a particular pattern.

Pattern Lab provides the ability to view both a pattern's underlying HTML and the template code used to generate the HTML. It can also be extended to showcase accompanying CSS and JavaScript code.

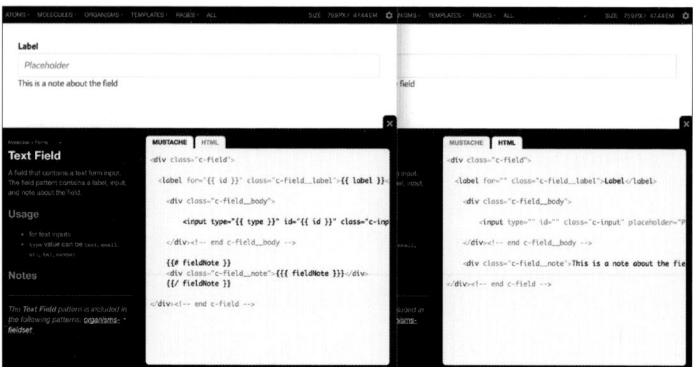

Pattern Lab's code view demonstrates both a pattern's template code and the compiled HTML.

Ultimately, whichever pattern library tool you decide to use should have some form of code view. Perhaps more importantly, the pattern libraries you create should showcase the types of code that enable you and your development team to be as effective as possible.

Living documentation and annotations

In a traditional siloed design process, it's typical to see lengthy wireframe and spec documents created, debated, and approved. These documents usually take the form of gigantic PDFs, which is unfortunate considering they often contain all sorts of valuable insights, instructions, and documentation about the design system. Sadly, these bulky artifacts are often thrown into a (virtual) trash can by the time the project makes its way into production.

This shouldn't be the case. A UI's documentation should contain insights from every discipline involved in creating it, and – this is key – should be baked into the living, breathing design system. Effective pattern libraries carve out a space to define and describe UI components, articulating considerations ranging from accessibility to performance to aesthetics and beyond.

Pattern Lab provides several ways to add pattern descriptions and annotations to a design system. Pattern descriptions can be added by creating a Markdown file that corresponds to the name of a pattern (e.g. *pattern-name.md*), which will show the pattern description in the library list view.

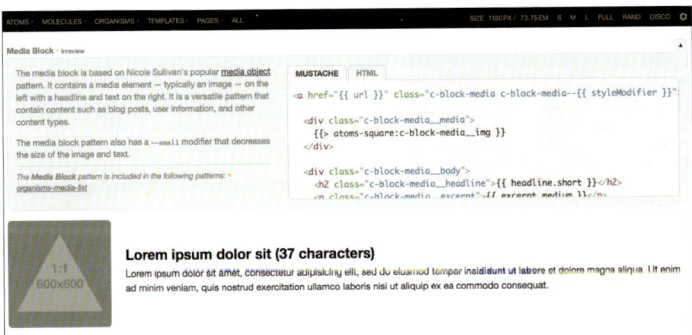

Pattern Lab displays important pattern documentation right alongside the living pattern examples, which helps teams communicate definitions, usage, examples, outside resources, and more.

Pattern Lab also provides a (dare I say) cool feature that enables you to attach annotations to any UI element and view those annotations in the context of the living, breathing design. When annotations are switched on, each annotated element receives a number which, when clicked, jumps you to the corresponding annotation. This allows teams to view pattern considerations within the context of the full UI. Pretty neat!

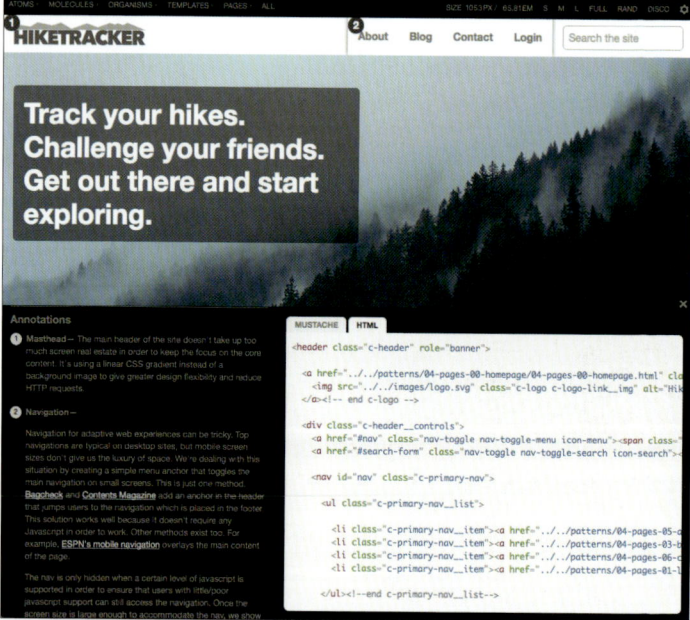

Pattern Lab's annotation feature is interactive and baked into the living UI.

Providing context with pattern lineage

When looking at various patterns in a library, I've found myself wondering, "Great, but where is this component actually used?" Defining and describing pattern characteristics is one thing, but there's an opportunity to provide additional contextual information about your UI patterns.

Thanks to the Russian nesting doll include approach described earlier, Pattern Lab can display what patterns make up any given component, and also show where those patterns are employed in the design system.

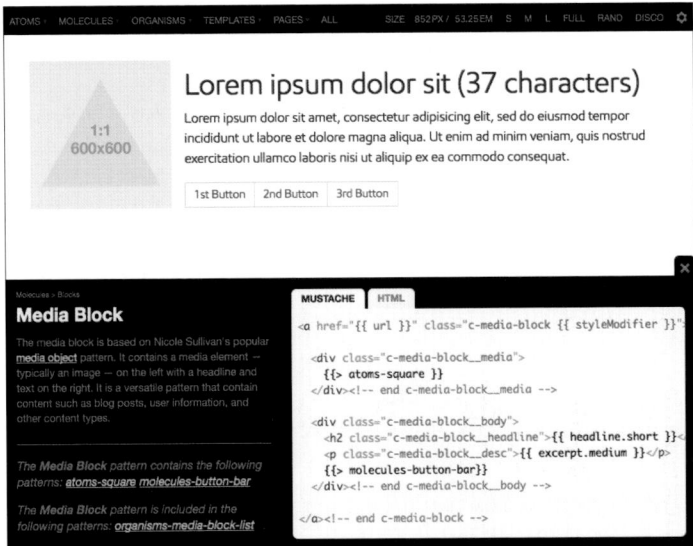

Pattern Lab's lineage feature displays what patterns make up any component, and also shows all the places that component is employed.

In the example above, we have a molecule pattern called `media-block`, which contains an image, headline, text, and a group of buttons. Looking at the pattern's lineage, we can see it contains a pattern called `atoms-square`, which is the thumbnail-sized image, as well as `molecules-button-bar`, which is the group of buttons. We can also see where exactly this pattern gets used: the `media-block-list` organism.

This contextual information is amazingly helpful for designers and developers; I know I use the lineage feature *all the time* in my own workflow. Let's say we wanted to make changes to a particular pattern, like doubling the size of an image or adding an additional element: **we'd immediately know which patterns and templates would need to be retested and QA'd to ensure nothing breaks with**

the changes. The lineage feature also helps point out redundant and underused patterns so teams can weed them out of the pattern library as the launch date gets closer.

To each their own

So there you have it. Pattern Lab provides several helpful features for teams to create deliberate, thoughtful design systems. But as I mentioned before, no single tool is going to be perfect for everyone and every situation. There are a ton of great tools[59] out there to help you create effective pattern libraries, and the tools you decide on will undoubtedly be influenced by your organization's environment, technologies, workflow, and culture.

When choosing tools to create your pattern library, you should keep your eyes open for these qualities and features of effective pattern libraries:

- Providing pattern descriptions and annotations.
- Showcasing the relevant pattern HTML, templating, CSS, and/or JavaScript code.
- Viewing patterns across the entire resolution spectrum.
- The ability to showcase pattern variations, such as active or disabled tabs.
- The ability to dynamically add real representative content into the patterns' structures.
- Providing contextual information, like which patterns make up a particular component, as well as where that component is used.

At the end of the day, it's not about the tools we use to create pattern libraries, but rather how we use them. Creating and maintaining an effective design system means dramatically changing your organization's culture, processes, and workflows. If that sounds hard to you, it's because it is. But fear not! The rest of the book will detail the entire process of creating and maintaining a successful design system to set your organization up for long-term success.

---

59 http://styleguides.io/tools.html

# Chapter 4
The Atomic Workflow

People, process, and making design systems happen

Talk is cheap. And up until now, we've been doing a whole lotta talkin'. That's not to say it hasn't been productive talk! After all, we've discussed the importance of modular thinking, we've learned a methodology for crafting deliberate UI design systems, and we've showcased tools for creating effective pattern libraries.

But here's where the rubber meets the road. Where we roll up our sleeves and put all of this theory into practice. Where we *get stuff done*. This chapter will tackle all that goes into selling, creating, and maintaining effective design systems. You ready? Let's go.

It's people!

The not-so-secret secret about creating effective design systems (or doing any great work, really): **it all comes down to people truly collaborating and communicating with one another.**

If this is such common knowledge, why aren't we constantly hearing thousands of success stories from around the world? I'll let Mark Boulton explain:

> *The design process is weird and complicated, because people are weird and complicated.*
>
> - Mark Boulton

You can have all the right technologies in place, use the latest and greatest tools, and even have extraordinarily talented individuals on board, but if everyone involved isn't actually cooperating and communicating with one another then you're not going to create great work. It's as simple as that. That's not to say you can't create *good* work, but more often than not you're going to create *one of the many disappointing shades of bad* work.

Establishing and maintaining successful interface design systems requires an organization-wide effort, and this chapter will discuss how to overcome human beings' many quirks to make them happen.

When to establish a design system

So when's the best time to establish an interface design system? Short answer: *now*.

Design systems and their accompanying pattern libraries are often created in conjunction with a new design or redesign project, replatforming effort, or other initiative. Piggybacking off another project is a great way to sneak a pattern library into your organization.

That being said, creating a design system and pattern library doesn't necessarily need to coincide with another project. If you can convince your clients and stakeholders to pony up the cash and resources for a dedicated design system project, then good on you!

How exactly do you sell a design system to your clients, bosses, colleagues, and stakeholders? Put on your business hat, because we're going to tackle that in the next section!

Pitching patterns

Introducing a new way of doing things is no easy task, as it requires changing people's existing mentalities and behaviors. Getting stakeholders and clients on board with establishing a design system involves constant education, in addition to a bit of marketing savvy.

First things first. It's necessary to introduce the concept of interface design systems to your clients, colleagues, and stakeholders. Explain what these design systems are and the many ways they can help the organization. We've discussed these benefits throughout the book, so you can explain how design systems **promote consistency and cohesion, speed up your team's productivity, establish a more collaborative workflow, establish a shared vocabulary, provide helpful documentation, make testing easier, and serve as a future-friendly foundation.** Who can say no to all that?!

Alas, I've found that I can hype design systems until I'm blue in the face, but the suits don't always see things through the same lens as the people on the ground. A simple reframing of the question helps immensely. It's much more effective to simply ask, "**Do you like saving time and money? Yes or no?**" If the answer to that question is *no*, I'm afraid you have way bigger problems than selling a design system. If the answer is *yes*, then you can pitch the benefits through the lens of time and money. Let's try this out, shall we?

- Design systems **lead to cohesive, consistent experiences.** That means users master your UI faster, leading to more conversions and **more money** based on the metrics your stakeholders care about.

- Design systems **speed up your team's workflow.** Rather than reinventing the wheel every time a new request comes through, teams can reuse already established UI puzzle pieces to roll out new pages and features faster than ever before.

- Centralizing UI components in a pattern library **establishes a shared vocabulary** for everyone in the organization, and **creates a more collaborative workflow** across all disciplines. With everyone speaking the same language, more time is spent getting work done and less time is spent dealing with superfluous back-and-forth communications and meetings.

- Design systems **make cross-browser/device, performance, and accessibility testing easier,** vastly speeding up production time and allowing teams to launch higher-quality work faster. Also, baking things like accessibility into a living design system scales those best practices, allowing your interfaces to reach more users while reducing the risk of you getting sued!

- Once a design system (with accompanying pattern library) is established, it **serves as a future-friendly foundation** for the organization to modify, tweak, extend, and improve on over time. Doing some A/B testing? Roll the lessons from those tests into the living design system. Made some big performance optimizations? Roll them into the living design system! The *living* part of living design systems means they can always adapt to meet the future needs of the organization, saving time and money all the while.

Framing things in terms of time and money helps the people controlling the purse strings understand why a design system is a worthwhile pursuit. With any luck, these conversations will translate to a concrete commitment from the organization (read: allocating real time and money) to make a design system happen.

Show, don't tell: the power of interface inventories

If only getting buy-in were as easy as having a few well-timed conversations with the right people. Maybe you're savvy enough to seal the deal with talking points alone, but for us mere mortals words aren't enough. Sometimes you need more. Sometimes you need to make them *feel the pain*.

Enter the interface inventory.

Many are familiar with the concept of a content inventory[60]. Content audits are usually performed in the early stages of a website redesign process to take stock of all a site's content. It's a tedious process involving spreadsheets and caffeine, but all that hard work pays off. By the end of the exercise the organization's content is laid out on the table, giving teams valuable insights into how to handle their content as they tackle the project.

An interface inventory is similar to a content inventory, only instead of sifting through and categorizing content, you're taking stock of and categorizing all the components that make up your user interface. **An interface inventory is a comprehensive collection of the bits and pieces that make up your user interface.**

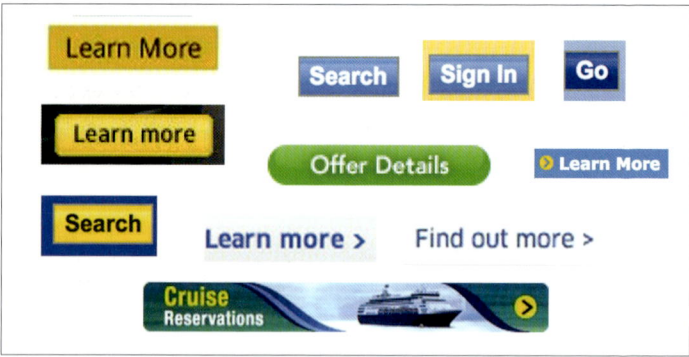

Here's a collection of all the unique button styles found on the homepage of United.com. An interface inventory rounds up and categorizes all the unique patterns that make up an interface.

60   https://en.wikipedia.org/wiki/Content_inventory

Conducting an interface audit

How do you go about conducting an interface audit? How do you round up all the components that make up your UI? The simple answer is *screenshots*. Lots of them! Creating an interface inventory requires screenshotting and loosely categorizing all the unique components that make up your user interfaces. While that's a relatively straightforward endeavor, there are some important considerations to keep in mind to make the inventory as useful as possible. Let's go through the process for creating a successful interface inventory.

Step 1: Round up the troops

I've encountered many ambitious designers and developers who have taken it upon themselves to start documenting their organization's UI patterns. While I certainly applaud this individual ambition, **it's absolutely essential to get all members of the team to experience the pain of an inconsistent UI for them to start thinking systematically.**

For the interface inventory to be as effective as possible, **representatives from all disciplines responsible for the success of the site should be in a room together** for the exercise. Round up the troops: UX designers, visual designers, front-end developers, back-end developers, copywriters, content strategists, project managers, business owners, QA, and any other stakeholders. The more the merrier! After all, one of the most crucial results of this exercise is to establish a shared vocabulary for everyone in the organization, and that requires input from the entire team.

Step 2: Prepare for screenshotting

The interface inventory exercise generates a ton of screenshots, so naturally you'll need software to capture and display those screenshots. Some possible tools include:

- PowerPoint or Keynote
- Photoshop or Sketch
- Evernote Web Clipper
- Google Docs or Microsoft Word

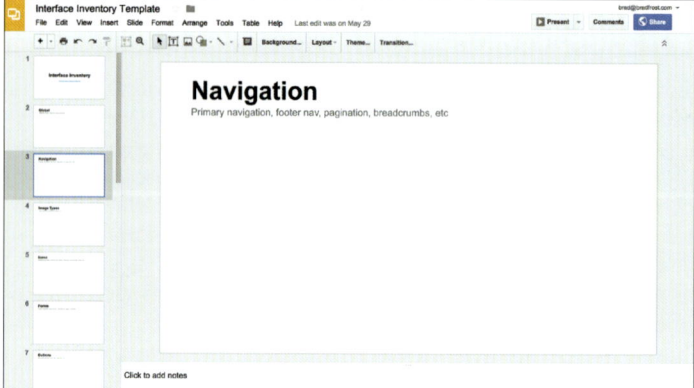

It's important for all participants to capture screenshots using the same software so they can be combined later. I've created a Google Slides interface inventory template for teams to use as a starting point.

- Google Slides (If you're interested, I've created a Google Slides interface inventory template[61])

Ultimately, it doesn't really matter what tool you use, but everyone should agree on a single tool to make it easier to combine at the end of the exercise. I've found online slide-building software like Google Slides to be very effective as it provides a canvas for free-form image positioning, they're chunked out into slides for easier categorization, and they're web-based so can be shared with ease.

Step 3: Screenshot exercise

Get your screenshotting fingers ready because it's time for the main event! **The interface audit exercise involves screenshotting and categorizing all the unique UI patterns that make up your experience.** Bear in mind this exercise doesn't mean capturing *every instance* of a particular UI pattern, but rather capturing *one instance of each* unique UI pattern.

Assign each participant a UI category. You may need to pair people or have participants document multiple categories, depending on how many people are taking part in the exercise. Once again, it's

---

61 https://docs.google.com/presentation/d/1GqFmiDV_NqKi36fXAwD3WTJL5-JV-gHL7XVD2fVeL0M/edit?usp=sharing

helpful to have as many participants as possible since more people screenshotting will result in more thorough documentation.

What patterns to capture

What interface element categories should be captured? Obviously, the categories are going to vary from interface to interface, but here are a few categories to start with:

- **Global elements:** components like headers, footers, and other global elements that are shared across the entire experience.
- **Navigation:** primary navigation, footer navigation, pagination, breadcrumbs, interactive component controls, and essentially anything that's used to navigate around a user interface.
- **Image types:** logos, hero images, avatars, thumbnails, backgrounds, and any other type of image pattern that shows up in the UI.
- **Icons:** icons are a special type of image worthy of their own category. Capture magnifying glasses, social icons, arrows, hamburgers, spinners, favicons, and every other interface icon.
- **Forms:** inputs, text areas, select menus, checkboxes, switches, radio buttons, sliders, and other forms of user input.
- **Buttons:** buttons are the quintessential UI element. Capture all the unique button patterns found throughout the experience: primary, secondary, big, small, disabled, active, loading, and even buttons that look like text links.
- **Headings**: h1, h2, h3, h4, h5, h6 and variations of typographic headings.
- **Blocks:** also known as touts, callouts, summaries, ads, or hero units, *blocks* are collections of typographic headings and/or images and/or summary text (see Nicole Sullivan's write-up about the _media object_[62] as an example of a block).
- **Lists:** unordered, ordered, definition, bulleted, numbered, lined, striped, or any group of elements presented in a list-type format.
- **Interactive components:** accordions, tabs, carousels, and other functional modules with moving parts.

[62] http://www.stubbornella.org/content/2010/06/25/the-media-object-saves-hundreds-of-lines-of-code/

- **Media:** video players, audio players and other rich media elements.
- **Third-party components:** widgets, iframes, stock tickers, social buttons, invisible tracking scripts[63], and anything else that isn't hosted on your domain.
- **Advertising:** all ad formats and dimensions.
- **Messaging:** alerts, success, errors, warnings, validation, loaders, popups, tooltips, and so on. This can be a challenging category to capture as messaging often requires user action to expose.
- **Colors:** capture all unique colors presented in the interface. This category can be aided by fantastic style guide bootstrapping tools like CSS Stats[64] and Stylify Me[65].
- **Animation:** animation is an elemental aspect of user interfaces, and should therefore be documented. This requires using screen recording software such as QuickTime to capture any UI element that moves, fades, shakes, transitions, or shimmies across the screen.

An example of unique button patterns captured in an interface inventory for a major bank's website.

63 http://bradfrost.com/blog/post/surfacing-invisible-elements/
64 http://cssstats.com/
65 http://stylifyme.com/

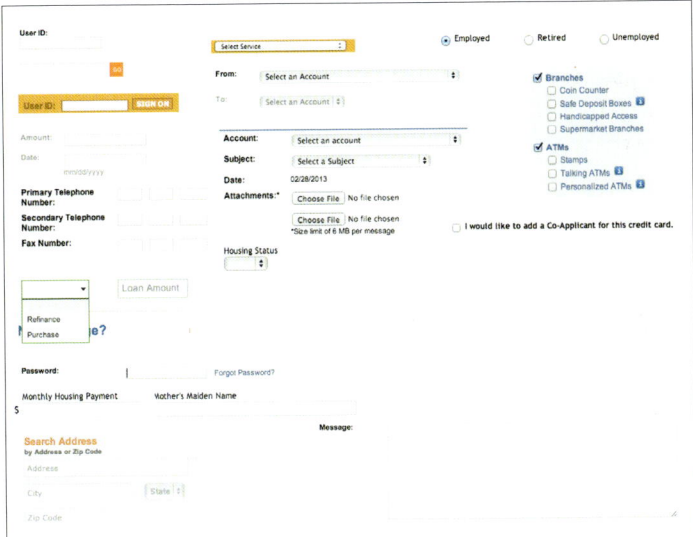

An example of various form elements captured in an interface inventory for a major bank's website.

Again, these categories aren't set in stone and will vary based on the nature of the user interface you're dealing with. Of course, it's important to add, subtract, or modify these categories to best fit your organizational needs.

Timing is everything

It's important to set time limits on the screenshotting exercise to avoid going down a rabbit hole that ends up lasting all day. The amount of time you allocate will vary depending on how many people are participating, but I find between **30 and 90 minutes** to be sufficient for a first pass of an interface inventory. Set a timer, throw on some Jeopardy! music (well, maybe not Jeopardy! music, but some other music that sets an upbeat mood for the exercise), and have the participants concentrate on screenshotting the unique UI patterns they encounter.

Dig deep

Which parts of the site should participants capture in the interface inventory? Short answer: *everything*. Any piece of UI that is or could be managed by your organization should be documented.

This is difficult as organizations tend to favor certain parts of the experience (*cough* homepage *cough*) over others. For example, people working on an e-commerce website tend to focus on the core shopping experience, even though areas like customer support, FAQs, sizing charts, 404 pages, and legal terms are also extremely important to the user experience. Remember, users perceive your brand as a singular entity and don't care about your organizational structure, tech stack, or anything else that might cause disparities in the UI. Encourage interface audit participants to be as thorough as possible during the exercise.

Step 4: Present findings

The screenshotting exercise can be a bit overwhelming, so be sure the team takes a break after the exercise is complete. Get some food, grab some coffee, and stretch your legs for a bit. Once everyone's feeling refreshed, it's time to discuss what you captured.

Have each participant spend five or ten minutes presenting each UI category to the group. Here's where the fun begins. Presenting to the group allows the team to discuss the rationale behind existing UI patterns, kick-starts a conversation about naming conventions, and gets the team excited to establish a more consistent interface.

Naming things is hard. It's fascinating to hear the inconsistent names designers, developers, product owners, and other stakeholders all have for the same UI pattern. "Oh, we call that the utility bar." "Oh, we call it the admin nav." "Oh, we call it the floating action area!" This exercise is an opportunity to unearth and iron out disparities between pattern labels, and also establish names for previously unlabeled patterns. Don't feel like you need to come to a consensus on patterns' final names in the course of ten minutes; this exercise is simply meant to open up a broader discussion.

Once every category has been presented and discussed, all the participants should send their slides to the exercise leader. The leader will then combine everything into one giant über-document, which will soon become a wrecking ball of truth and justice.

Step 5: Regroup and establish next steps

With the über-document in hand, it's time to get the entire organization on board with crafting an interface design system.

Have you ever wanted to see a CEO cry? Laying bare all your UI's inconsistencies is a great way to make that happen! **One of the most powerful benefits of interface inventories is that you can show them to anyone, including non-designers and developers, and they'll understand why inconsistent UIs are problematic.** You don't need to be a designer to recognize that having 37 unique button styles probably isn't a good idea. Here's your opportunity to make it crystal clear to stakeholders that approaching your UI in a more systematic way makes great sense for both your users and your organization.

In addition to selling the idea to key stakeholders, **all the hard work and discussion that went into the initial interface inventory exercise should be translated into the seeds of your future design system and pattern library.**

It's very likely the initial exercise didn't capture every unique UI pattern, so you may need to conduct another interface audit exercise to capture a more complete picture of your UI patterns. This may involve a large group again, but in reality a smaller, cross-disciplinary team will be going through the über-document and establishing next steps for the design system.

Once the gaps in the interface inventory have been filled, the working group can have some important conversations about next steps for the design system project. Some key questions for this group to cover include:

- Which patterns should stay, which should go, and which can be merged together?
- What pattern names should we settle on?
- What are the next steps to translate the interface inventory into a living pattern library?

Benefits of an interface inventory

Creating an interface inventory can be quite an undertaking, but the benefits of making one are many:

- **Captures all patterns and their inconsistencies:** an interface inventory rounds up all the unique patterns that make up your UI. Seeing all those similar, but still different, patterns next to each other exposes redundancy and underscores the need to create a consistent, cohesive experience.

- **Gets organizational buy-in:** having a large, diverse group of disciplines participate in the exercise helps everyone understand the value of creating and maintaining a consistent user interface. Also, the interface inventory über-document can be a tremendously powerful tool for convincing stakeholders, bosses, and clients to invest in an interface design system.

- **Establishes a scope of work:** an interface inventory helps design teams determine the level of effort required to design and build each UI pattern as part of a design or redesign project. Which components will be relatively easy or difficult to convert into a responsive environment? What are the content, design, and development considerations around each component? An interface inventory enables teams to have important conversations that help establish a project's realistic scope and timeline.

- **Lays the groundwork to a sound interface design system:** the interface inventory is an important first step for setting up a comprehensive pattern library. It's essential to capture all existing UI patterns to determine which patterns will make the final cut in the living design system. The interface audit exercise also helps teams establish a shared vocabulary, which will be crucial for the success of the eventual design system.

Ask forgiveness, not permission

So you've discussed the benefits of establishing a living design system with your stakeholders, and you've even created an interface inventory to show them the inconsistent train wreck that is the current UI. And yet, despite all your efforts, they still shoot down the sound idea of establishing an interface design system and pattern library. What's a responsible web team to do?

Do it anyways.

Just how we build things like performance, accessibility, and responsiveness into our products and process by default, we should also create design systems by default. You don't need to get the client's blessing to follow your craft's best practices. When you give stakeholders the option to say no to something, they will. So simply don't give them that opportunity. **Our job is to create great work for our clients and organizations, and interface design systems are a means to that end.**

In fact, **to create the whole, you need to create the parts of that whole.** Our interfaces consist of smaller pieces, whether you pay those smaller pieces any mind or not.

You have a decision to make: focus solely on creating the whole while ignoring the parts, or spend some time organizing the parts to help you more efficiently create the whole. In his book _Multiscreen UX Design_[66], Wolfram Nagel wonderfully articulates these approaches using Lego bricks as an analogy.

One way to approach a Lego project is to simply dump the pieces out of the box onto a table, roll up your sleeves, then start building your creation.

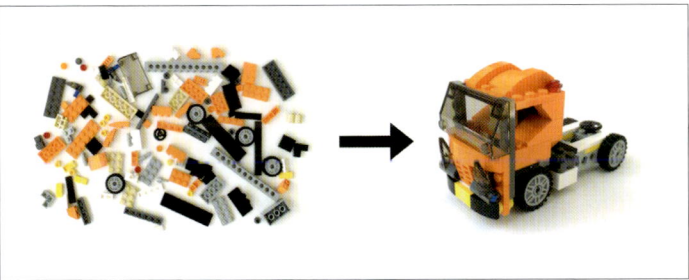

One way to approach a Lego project is to simply dump the pieces out onto a table, and rummage through the pile to find the pieces you need. Image adapted from "Multiscreen UX Design" by Wolfram Nagel.

---

[66] http://store.elsevier.com/Multiscreen-UX-Design/Wolfram-Nagel/isbn-9780128027295/

This approach to a Lego project is certainly a viable strategy, even if it is unapologetically haphazard. The only time you'd pay attention to the pile of bricks is when you're sifting through it to find the specific piece you need.

This is not dissimilar to how many digital projects are approached. The client needs a website, so we jump in to designing and building it. The client needs a mobile app, so we immediately start building the screens of the app. Our gaze remains transfixed on the final product, and we rarely, if ever, glance at the underlying patterns that comprise our final UIs.

Of course, there is another way to approach your Lego and digital projects. Rather than diving headfirst into constructing the final work, you can take the time to take stock of the available pieces and organize them in such a way that they become more useful.

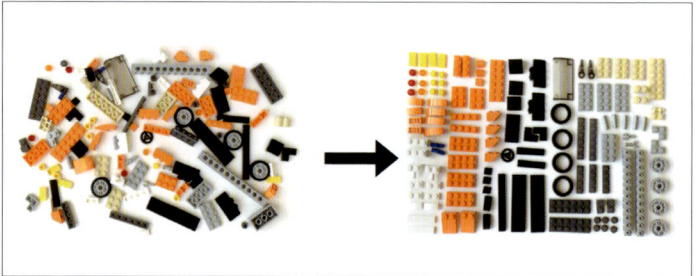

Taking the time to organize the pieces that make up your final creations allows you to approach the work in a more deliberate and efficient manner. Image adapted from "Multiscreen UX Design" by Wolfram Nagel.

No doubt organizing takes time, planning, and effort. It doesn't come for free. The fact that this configuring isn't visibly represented in the final product may tempt us to say it serves as a distraction to the real work that needs to be done. Why bother?

By taking the time to organize the parts, you can now create the whole in a more realistic, deliberate, and efficient manner. Creating a library of your available materials allows you to approach the project in a more methodical way, and saves immense amounts of time in the process. Rather than rummaging through a haphazard pile of bricks and burning time reinventing patterns, you can create an organized system of components that will help produce better work in a shorter amount of time.

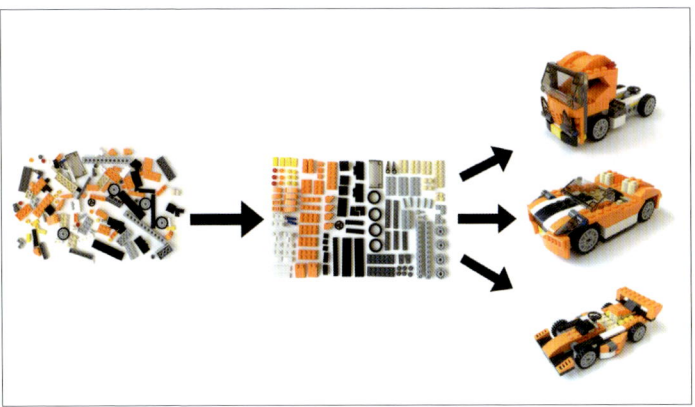

Taking the time to organize the pieces that make up your final creations allows you to work in a more deliberate and efficient manner. Rather than sifting through a haphazard pile of bricks, an organized inventory of components can produce better, faster work. Image adapted from "Multiscreen UX Design" by Wolfram Nagel.

As far as your clients and stakeholders are concerned, the final product is still being produced. So long as you're showing progress on the final work, you can decide how much of your internal process you're willing to expose. The fact that you're creating a design system to produce the final product is really of no concern to them; it's simply a decision your team is making to create better work.

If you're dealing with change-averse stakeholders, I say do what you need to do and tell them to pay no attention to what's happening behind the scenes. Once you've successfully launched the project and the champagne has been poured, you can pull back the curtain and say, "Oh, by the way, we established a design system and pattern library so the team could collaborate and work more efficiently together." It would be extremely difficult for them to argue against you now, especially if the project came in on time and on budget. If you're really lucky, you can parlay the initial project's success into a more official initiative within the organization to evolve your design system.

Of course, it's preferable to get your clients, colleagues, and stakeholders excited about creating an interface design system, or at the very least get their blessing to pursue the project in a modular fashion. But I think it's important to find ways to follow your craft's best practices even when you're faced with extreme organizational resistance.

(Re)setting expectations

You've put in a lot of hard work to sell the concept of a design system, but you still need to set stakeholder and team expectations before you roll up your sleeves and get to work.

When I say "set expectations" I'm really saying "**re**set expectations." You see, we all bring our own experiences, opinions, and biases to a project. Our industry is still incredibly young, so many people working on web projects are coming from other industries with their own established Ways Of Doing Things™. Even people who have worked exclusively in the digital world have felt the baggage of industries past. Moreover, the guiding principles, best practices, and tactics of digital design are still very much being codified.

It's ludicrous for anyone to utter the phrase, "This is how we've always done things" in an industry that's only 25 years old. Unfortunately, we humans are creatures of habit, and stepping outside familiarity's warm embrace is uncomfortable. We don't like being uncomfortable. We must overcome our existing predispositions if we're going to embrace our ever-shifting industry's best practices and create successful digital work.

Redefining design

We've come a long way from simply transplanting print PDFs to the World Wide Web, but **print design still casts a long shadow and continues to influence how things get done online.**

Design in the print world focuses heavily on visual aesthetics. After all, you can't do much more with a poster than look at it. To be clear, I'm certainly not implying print design is easy or one-dimensional; the world of print is steeped in nuance and craft. What I am saying is that **the bidirectional and interactive nature of the web adds many more dimensions to what constitutes good design.** Speed, screen size, environment, technological capabilities, form-factor, ergonomics, usability, accessibility, context, and user preferences must be considered if we want to create great work for this brave new digital world.

These additional design considerations are vital for creating great digital work, yet they are too often absent from our processes and workflows. Designer Dan Mall explains:

> As an industry, we sell websites like paintings. Instead, we should be selling beautiful and easy access to content, agnostic of device, screen size, or context.
>
> - Dan Mall[67]

How did we get to the point where we sell and design websites like they're static images? During the formative years of the web we created experiences meant to be consumed solely by desktop computers, which is understandable since desktops were really the only game in town. The real estate provided by desktop screens made the idea of simply translating a PDF onto the web feasible and enticing. So that's what we did – and for a while it actually worked!

Once upon a time, the web was primarily consumed on desktop screens, hence this crusty-looking, old machine.

---

67 http://danielmall.com/articles/the-post-psd-era/

However, this didn't come without consequences. This print-like perspective of the web reinforced the notion that web designs, like their offline counterparts, could and should look the same in every environment. It also kept the focus on how a web design *looked* rather than how it *worked*, ignoring all the unique characteristics of this rich new medium. Moreover, it strengthened the belief that we could apply the same linear processes used to create print work to our digital work.

Time went by, of course, and mobile exploded, technology improved, and the web become the incredibly large and diverse landscape we know today. Gone are the desktop-only days of yore, and in their place is a time of smartphones, dumb phones, tablets, phablets, netbooks, notebooks, e-readers, wearables, TVs, game consoles, car dashboards, and so much more.

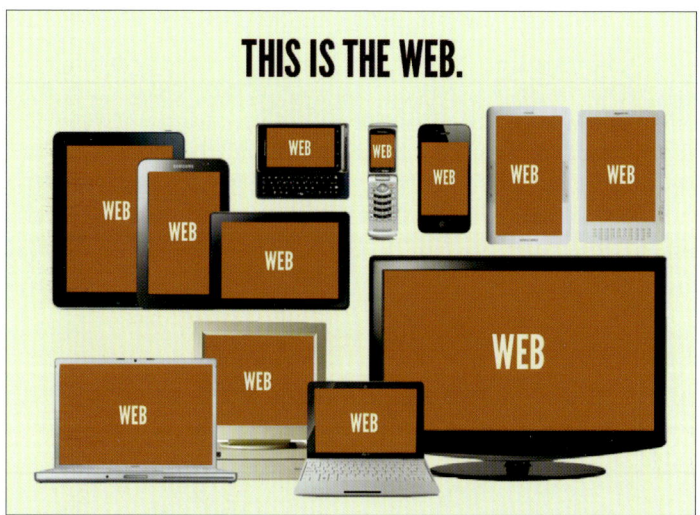

This is the web: a potpourri of devices, screen sizes, capabilities, form factors, network speeds, input types, and more.

The diversity of today's web landscape has shattered the consensual hallucination[68] of the desktop web, where we could simply bolt on the mentalities and processes of print to this new medium. Simply

68 https://adactio.com/journal/4443

looking at a smartphone, tablet, and desktop machine next to one another quickly erodes the assumption that a web design should look the same in every environment.

We're still at the very beginning of the Big Bang of connected devices. The device and web landscape of tomorrow will undoubtedly be even bigger and diverse than today's. In addition to current devices and the nascent technologies already on the horizon, the future web will involve technologies and ideas that haven't yet been conceived.

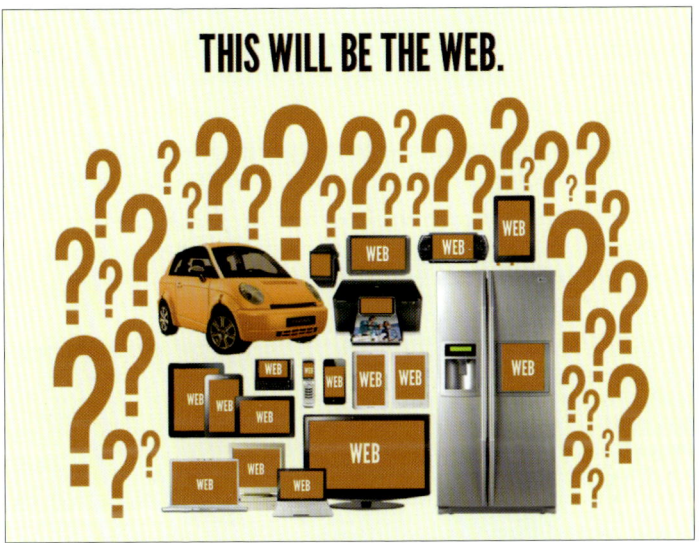

In addition to all the web-capable devices we concern ourselves with today, we must understand that the device and web landscape is becoming bigger and more diverse all the time.

I've found the three previous images to be a tremendously helpful shorthand for helping clients, colleagues, and stakeholders understand the reality of the web landscape. With this newfound understanding, everyone becomes a whole lot more receptive to updating their processes and workflows to create great work for this unique medium.

It's our job to create great experiences for people using a diversity of devices, screen sizes, network speeds, device capabilities, browser features, input types, form factors, contexts, and preferences. That's undoubtedly a Herculean task, but all these variables really underscore the need to extend far beyond visual aesthetics when creating interface design systems.

In addition to making visually beautiful and consistent experiences, we should:

- **Embrace the ubiquity of the web** by creating accessible, resilient design systems. Recognize that a whole slew of people with a vast spectrum of capabilities will be accessing our experiences, so construct design systems to be as inclusive as possible.
- **Create flexible layouts and components** so our interfaces look and function beautifully irrespective of any particular device dimension or screen size.
- **Treat performance as an essential design principle** and create fast-loading experiences that respect users and their time.
- **Progressively enhance our interfaces** by establishing core experiences then layering on enhancements to take advantage of the unique capabilities of modern devices and browsers.
- **Create future-friendly design systems** meant to stand the test of time and anticipate inevitable changes to the device and web landscape.

Of course, there are many other design considerations that should be included in our interface design systems (ergonomics, input type, Section 508 compliance, legibility, and so on), but the key takeaway here is to expand the definition of what constitutes good digital design beyond visual aesthetics.

As you might expect, substantial changes to our processes need to happen so we can properly address all these uniquely digital design considerations. It therefore becomes our responsibility to set client, colleague, and stakeholder expectations so that everyone knows the process for creating will be different this time around.

Death to the waterfall

Tell me if you've heard this one before. A team is tasked with making a website. Once the kick-off meeting dust has settled, a UX designer goes away, puts their head down, and eventually emerges with a giant PDF document detailing the entire experience. This monolithic wireframe document gets passed around to the project stakeholders, who sign it off after some feedback and suggestions.

The UX designer then passes the wireframes to the visual designer, who hops into Photoshop or Sketch to apply color, typography, and texture to the structured-but-sterile wireframes. In the design review meeting, stakeholders sit eagerly while the projector fires up and the project manager runs off to print copies of the design deck for everyone. The art director takes their position at the front of the room and unveils the design. Behold, a website design! Once the presentation is finished, the room quickly buzzes with feedback and conversation. After the initial reactions and compliments die down, a key stakeholder speaks up.

"This looks fantastic, and I think really hits the mark for what we're trying to accomplish with this project. *But...*"

They express their desire to see something perhaps with an alternate layout, something that captures a certain vibe, maybe something that uses different photography, something that just... *pops.*

With the floodgates opened, the other stakeholders suddenly realize they too have opinions and constructive criticism they'd like to share. By the time the meeting draws to a close, everyone has rambled off their wish list of what they'd like the design to accomplish.

Slightly deflated but determined to nail it, the visual designer retreats back to their tools to work in the stakeholders' suggestions. At the next design review meeting, the same scene repeats itself, with stakeholders expressing equal parts encouragement and longing for more. "I feel like we're almost there. Could we just..."

Weeks pass and seasons change. Nerves wear thin, and the deadline date looms over everyone's heads. It's with a sense of urgency that *homepage_v9_final_for-review_FINAL_bradEdits_for-handoff.psd* finally gets approval by the stakeholders.

The visual designer, relieved they've finally completed their job, tiptoes oh-so-quietly up to the entrance of the Code Cave. They slip the approved design under the door, and as they scamper away they yell, "Can you get this done in three weeks? We're already behind schedule and we're out of budget!"

The visual designer has already disappeared into the night by the time the front-end developer picks the design off the floor. With one glance at the composition, a strange feeling – some combination of bewilderment, rage, and dread – washes over them. What's wrong with the design, exactly? Maybe it's the seven typefaces and nine unique button styles peppered throughout the comps. Maybe it's the desktop-centric, impossible-to-actually-execute layout. Maybe it's the perfect-yet-improbable user-generated content.

The front-end developer tries in vain to raise their concerns to the broader group, but is quickly dismissed as being either inept or curmudgeonly. Alas, it's too late in the game to make significant changes to the design, especially since it's already been approved by the stakeholders.

So the developer tries their best to make lemonade out of the lemony static comps. They bend over backwards to create responsive layouts that still retain the integrity of the static comps, normalize some of the more blatant component inconsistencies, establish pattern states (like button hover, active, and disabled states) that weren't articulated in the designs, and make some on-the-fly decisions regarding the interactive aspects of the experience. Discussions with designers are strained, but everyone realizes that they need to work through these issues to get the project done.

After plugging the front-end code into a CMS, frantically finalizing the site's content, and doing some last-minute QA testing, the team finally launches the site. While no one says it out loud, there's a tinge of disappointment in the air alongside the joy and relief of getting the project out the door. After all, the live site lacks the glossy polish that the comps promised to the stakeholders, and friction between disciplines has bruised some relationships.

I hope this story reads as a work of fiction to you, but based on my own experiences and conversations with countless others, I'm guessing you've experienced this tale of woe at one point or another. It may even hit home like a punch in the gut. Whether you've endured this process firsthand or not, it's important to recognize that the Henry Ford-esque waterfall process[69] increasingly isn't likely to result in great digital work.

69   https://en.wikipedia.org/wiki/Waterfall_model

The waterfall process, where disciplines pass off work to each other in sequential order, isn't likely to result in great digital work.

The waterfall process may make sense for print, architecture, manufacturing, and other physical media since mistakes and changes are extraordinarily costly. If a team overlooks an error made early in the process, they'll pay dearly for it later. However, **the digital world isn't constrained by the same limitations as the physical one. Pixels are cheap.** Changes can happen in an instant, hypotheses can be quickly tested out, and designs and code can be iterated on again and again.

The waterfall process hinges on the premise that work must flow in a sequential order: the UX designer's work must be completed before visual design can start; the visual designer must finish their job before front-end development can begin. This simply isn't true. There is much work that can and should happen in parallel. To create sound UI design systems, we must reset our stakeholders' expectations and get them comfortable with a blurrier, more collaborative process.

That work will happen in parallel doesn't imply that everyone will be guns blazing throughout the entire process. Of course, the bulk of research, information architecture, and other elemental aspects of UX design will tend to happen earlier in the process, but that work shouldn't delay the other disciplines from starting their jobs. And even when the bulk of a person's *active* work is done, they should never simply fade away from the project. It's crucial for every discipline to continue to consult with the others to ensure their vision makes it into the final product. So rather than a rigid, sequential waterfall process, a more collaborative process over time looks something like this:

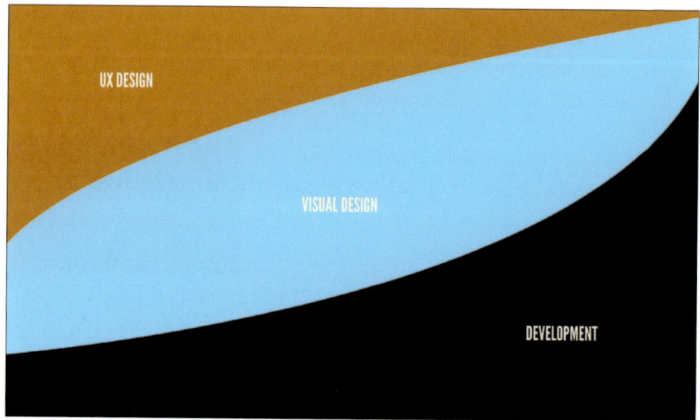

A more collaborative workflow involves a cross-disciplinary team working together throughout the entire process. While active work will wax and wane, each discipline continues to consult with the other team members to ensure their insights are present in the final work.

Development is design

When a previous employer discovered I wrote HTML, CSS, and presentational JavaScript, they moved me to sit with the engineers and back-end developers. Before too long I was being asked, "Hey, Brad. How long is that middleware going to take to build?" and "Can you normalize this database real quick?"

Here's the thing: I've never had a computer science class in my life, and I spent my high school career hanging out in the art room. Suffice it to say those requests made me extremely uncomfortable.

There's a fundamental misunderstanding that all coding is ultra-geeky programming, which simply isn't the case. HTML is not a programming language. CSS is not a programming language. **But because HTML and CSS are still code, front-end development is**

often put in the same bucket as Python, Java, PHP, Ruby, C++, and other programming languages. This misunderstanding tends to give many front-end developers, myself included, a severe identity crisis.

Organizationally, there is often a massive divide between designers and developers (or marketing and IT, or creative and engineering, or some other divisive labels). Designers and developers often sit on different floors, in different buildings altogether, in different cities, and sometimes even in different countries on different continents. While some of this organizational separation may be justified, **creating a division between designers and front-end developers is an absolutely terrible idea.**

The fact remains that HTML, CSS, and presentational JavaScript build user interfaces – yes, the same user interfaces that those designers are meticulously crafting in tools like Photoshop and Sketch. For teams to build successful user interface design systems together, **it's crucial to treat front-end development as a core part of the design process**[70].

When you show stakeholders only static pictures of websites, they can naturally only comment and sign off on pictures of websites. This sets the wrong expectations[71]. But by **getting the design into the browser as fast as possible**, you confront stakeholders with the realities of the final medium much sooner in the process. Working in HTML, CSS, and presentational JavaScript allows teams to not only create aesthetically beautiful designs, but demonstrates those uniquely digital design considerations like:

- flexibility
- impact of the network
- interaction
- motion
- ergonomics
- color and text rendering
- pixel density

---

70 http://bradfrost.com/blog/post/development-is-design
71 https://stuffandnonsense.co.uk/blog/about/time_to_stop_showing_clients_static_design_visuals

- scrolling performance
- device and browser quirks
- user preferences

Crucially, jumping into the browser faster also kick-starts the creation of the patterns that will make up the living, breathing design system. More on this in a bit.

This is not to say teams must design *entirely* in the browser. As with anything, it's about using the right tools at the right time to articulate the right things. Having the design represented in the browser *in addition to* other design artifacts gives teams the ability to paint a richer, more realistic picture of the UI they're crafting. Teams may demonstrate an aesthetically focused design idea as a static image, and simultaneously demonstrate a working prototype of that same idea in the browser.

An iterative iterative iterative iterative process

I believe a successful digital design process is quite similar to subtractive stone sculpture. At the beginning of the sculpting process, the artist and their patron have a general idea of what's being created, but that vision won't be fully realized until the sculpture is complete.

The sculptor starts with a giant slab of rock and starts chipping away. A crude shape begins to form after the first pass, and the shape becomes more pronounced with every subsequent pass. After a few rounds of whacking away at the rock, it becomes clear that the sculptor's subject is a human form.

With the general shape of the sculpture roughed out, the artist then begins homing in on specific sections of the piece. For instance, they may begin with the face, moving up close to carve the shape of the eyes, nose, and mouth. After several passes, they then move on to the arms, and then begin detailing the legs. At regular intervals the artist steps back to see how their detailed work affects the overall sculpture. This process continues until the sculpture is complete and everyone is pleased with the results.

Again, I think subtractive stone sculpture is a great analogy for a successful digital process, although unlike sculpture we have the power of *undo*!

An iterative digital process is similar to subtractive stone sculpture, where fidelity is built up over many iterations. Image credit: Mike Beauregard on Flickr[72]

It's essential to get stakeholders comfortable with reviewing works in progress rather than fully baked designs and code. As I mentioned in chapter 1, every organization these days wants to become more agile, and iteration is a key part of being agile. **It's more important to make steps in the right direction** than exhaust a ton of effort painting unrealistic pictures of what you want the final piece to be. **A sound design system doesn't roll off an assembly line, but is rather sculpted in iterative loops, building up fidelity as the project progresses.**

If this all sounds a bit messy, that's because it is! To the dismay of some project managers, the design process doesn't fit neatly into the rigid borders of Excel spreadsheets and Gantt charts. True collaboration between disciplines is fuzzy and chaotic, and that's not a bad thing. **Constant communication, tight feedback loops, and true collaboration therefore become the glue that holds**

[72] https://flic.kr/p/dLrf6w

the process together. Get your entire team to commit to honest conversation and genuine collaboration, and the details of your process will fall into place.

Are everyone's expectations properly set? Good! Now let's roll up our sleeves and get to work establishing our design system.

Establishing direction

Teams are often eager to jump right into fun high-fidelity design and development work, and clients are eager to see and react to that detailed work. However, this leads to distractions, assumptions, and all the aforementioned misguided expectations. **It's essential to agree on an overall design direction and paint the broad strokes first** before moving into high-fidelity design and development. This requires restraint and expectation management, but results in more focused decision-making and more realistic work.

What does this lo-fi work look like? Let's take a look at some techniques UX designers, visual designers, and front-end developers can use to begin crafting a strong overall direction for a UI design system.

Establishing content and display patterns

There's a ton of up-front strategic and research work that can and should happen toward the beginning of a project. *UX designers* (known by other monikers such as *information designers*, *information architects*, *interaction designers*, and so on) are responsible for synthesizing all that vital information and translating it into a user interface that meets the project's business and user goals.

In a traditional waterfall process, many UX designers have gone about this task by generating high-fidelity wireframes that document every screen of the entire user experience. These wireframe documents, stuffed to the gills with black rectangles and annotations, spec out the details of what the interface will accomplish, and are used to get stakeholder buy-in. As thorough as these documents tend to be, they don't paint the full picture and often make dangerous assumptions about visual layout and technical functionality.

Rather than jumping straight into such high-fidelity documents, it's better to **start with lo-fi sketches that establish what appears on a particular screen and in what general order**. Establishing the experience's basic information architecture can be accomplished with a simple bulleted list and a conversation. For a project I did for the Greater Pittsburgh Community Food Bank, I started by stubbing out the basic information architecture for a page on a site.

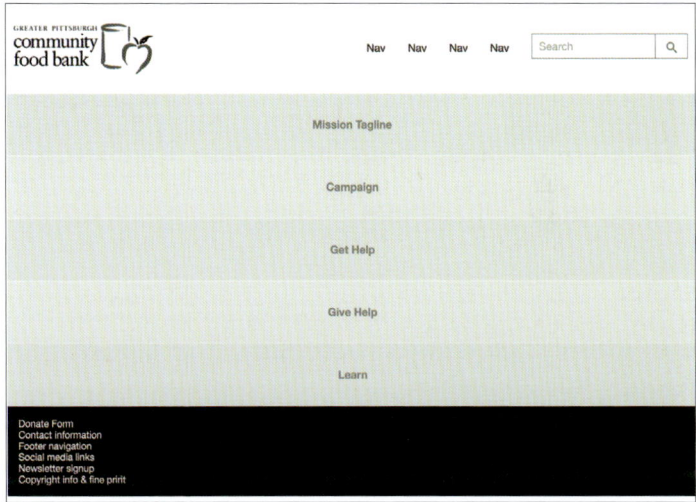

Basic HTML wireframes for the Greater Pittsburgh Community Food Bank homepage.

No one in their right mind would mistake this blocked out grayscale page as complete, but it provides more than enough information to have important conversations about the page structure and hierarchy.

Making lo-fi wireframes *mobile-first*[73] means using the constraints of small screens to force the team to focus on the core content and hierarchy. You can now ask, "Do we have the right things on this screen?" "Are they in the right general order?"

---

73 http://www.lukew.com/ff/entry.asp?933

These blocky grayscale wireframes help establish the necessary content patterns[74] for the screen, but UX designers can also articulate some site-wide UI patterns they anticipate using to ultimately display those content patterns. For the redesign of TechCrunch, designer Jennifer Brook defined a few site-wide UI patterns that could be used anywhere:

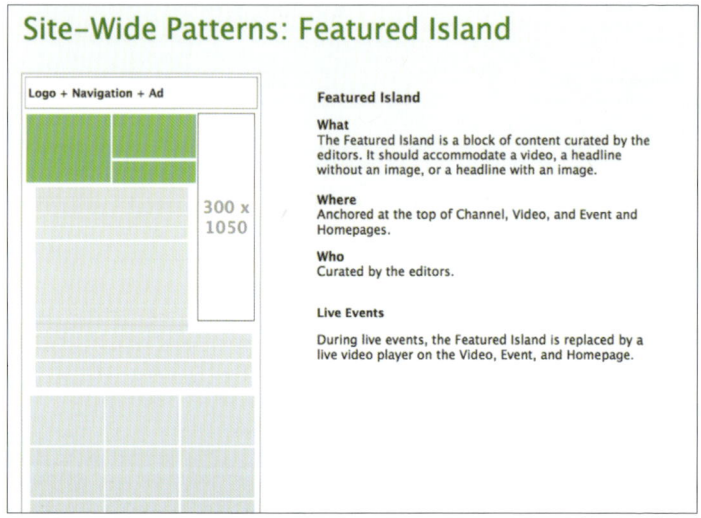

For the TechCrunch website redesign, Jennifer Brook defined site-wide, gestural display patterns, which don't make assumptions about aesthetics or functionality.

From the image above, you can gather that the "featured island" component will show content in some fashion. Note the gestural nature of this sketch and how it doesn't make any specific assumptions about layout or functionality. The details of how this pattern will look and function will come later, but at the beginning of the project it's useful simply to define it and articulate where it might get used.

As I've discovered from subsequent projects, content and display patterns can be effectively communicated in an even simpler format: the lowly spreadsheet.

---

74 http://danielmall.com/articles/content-display-patterns/

A simple spreadsheet can articulate what content and display patterns go on a given page while describing their order and purpose.

With a few simple spreadsheet columns, we can articulate which display patterns should be included in a given template, and what content patterns they'll contain. More importantly, we're able to articulate each pattern's relative hierarchy and the role it plays on the screen. If you read the leftmost column vertically, you're effectively looking at the mobile-first view of what the UI could be.

"What content and display patterns go on this page? And in what general order?" are crucial questions to ask, and the techniques we just described can help designers discuss them effectively without making any layout or technical assumptions.

Establishing visual direction

A visual designer's job is to create an aesthetic language and apply it to the user interface in a way that aligns with the project's goals. To do this, it's essential for a visual designer to unearth the stakeholders' aesthetic values.

Historically, visual designers have gone about this by creating full comps – often *many* comps – to feel out the aesthetic values of the organization. Throw some comps against the wall and see what sticks. As you might imagine, generating a slew of comps from scratch takes an immense amount of time and effort, and unfortunately much of that work finds itself on the cutting room floor. There must be a more efficient way.

As it turns out, there's a better path to take to arrive at aesthetic values without having to do a hell of a lot of up-front design work. Let's talk about some of the tactics for making this happen.

The 20-second gut test

A fantastic exercise for quickly establishing aesthetic values is the 20-second gut test[75]. Typically done as part of the project kick-off meeting, the exercise involves showing the stakeholders a handful of pertinent websites (about twenty to thirty of them) for twenty seconds each. The sites you choose should be a healthy blend of industry-specific sites and other visually interesting sites from other industries. For added believability, you can photoshop in your client's logo in place of the site's actual logo.

For each site presented, each person votes on a scale from 1 to 10, where a score of 1 means "If this were our site I would quit my job and cry myself to sleep," while a score of 10 means "If this were our site I would be absolutely ecstatic!" Instruct participants to consider visual properties they find interesting, such as typography, color, density, layout, illustration style, and general vibe.

For the Pittsburgh Food Bank website redesign kick-off, we showed stakeholders a variety of relevant websites for twenty seconds each. The participants voted on how happy they would be if the particular site was theirs. Then we discussed the results.

When the exercise is complete, quickly tally up the scores and come back to the group to discuss the results. Have a conversation about the sites that received the five lowest scores, the five highest scores, and the most contentious scores (sites which some people ranked very high and others ranked very low). The participants should explain why they were attracted or repulsed by a particular site, and work through differences in opinions with the group.

This exercise exposes stakeholders to a variety of aesthetic directions early in the process, allows them to work through differences in taste, and (with any luck) helps arrive at some shared aesthetic values. The visual designer can then latch on to these insights and begin to translate those aesthetic values into a visual direction for the project.

Style tiles

Once again, visual designers' first instinct is often to jump right into creating full comps to articulate an aesthetic direction for the project. This high-fidelity work is certainly tangible, but also wastes a ton of time and effort if the comps don't resonate with the stakeholders. Also, creating high-fidelity comps often makes big assumptions about technical feasibility, which leads to unrealistic expectations and antagonistic relationships with front-end developers.

It's essential to establish a solid visual direction for the project, so how does a visual designer do that without burning a ton of time on up-front high-fidelity comps? That's the question that designer Samantha Warren answered when she created style tiles[76], a deliverable that's more tangible than a mood board but not as high-fidelity as a fully baked comp.

---

75   http://goodkickoffmeetings.com/2010/04/the-20-second-gut-test/
76   http://styletil.es/

For the Entertainment Weekly website redesign project, visual designers used style tiles to explore color, type, texture, and more.

Style tiles (along with their in-browser counterparts, style prototypes[77]) allow designers to explore color, typography, texture, icons, and other aspects of design atmosphere without making assumptions about layout or worrying about polish. They can be designed much faster because they're not encumbered by the expectations of high-fidelity comps, which means feedback and discussion can happen sooner.

77 http://sparkbox.github.io/style-prototype/

Style tiles facilitate conversation to uncover what stakeholders value and what they don't. "Does this style tile resonate better with you rather than this one? Why?" "Why does this color palette not sit well with you?" "What is it about this typeface you like?" You can have important conversations about aesthetic design without having to create full comps.

Crucially, style tiles also reinforce pattern-based thinking by educating stakeholders about design *systems* rather than *pages*. Presenting color swatches, type examples, and textures exposes stakeholders to the ingredients that will underpin any implementation of the design system.

Element collages

While style tiles are great for exploring design atmosphere, they're still a bit abstract. To get a sense of how those design ingredients will be applied to an interface, it's important to quickly move into something a bit more tangible than a style tile. But does that mean visual designers need to jump from style tiles straight into full comps? Not necessarily.

Somewhere in between style tiles and full comps live element collages[78], which are collections of UI component design explorations. Element collages provide a playground for designers to apply design atmosphere to actual interface elements, yet still be free from layout and highly polished presentation.

---

78 http://danielmall.com/articles/rif-element-collages/

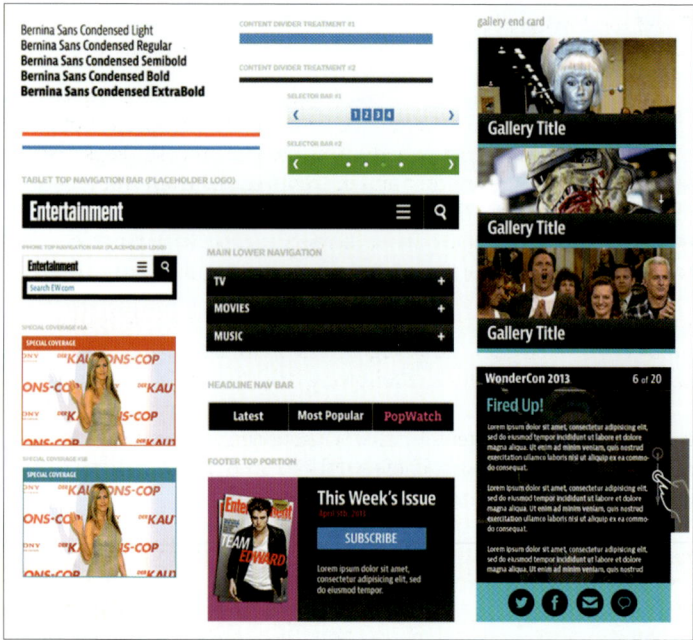

An element collage for the Entertainment Weekly redesign applied color, typography, and texture to actual interface elements. These collages enabled important conversations about the aesthetic direction of the project.

Like style tiles, element collages are meant to facilitate discussion about the aesthetic direction of the project. It's very clear these collages aren't an actual website, but stakeholders can still get a sense of what the site *could* look like. Conversation about these element collages can give visual designers more ideas and direction about where to take the design next, and because of their lo-fi nature, designers can quickly iterate and evolve ideas.

No doubt other tactics exist to establish aesthetic direction for your projects, and which techniques you decide to employ will vary from project to project. But the key is to paint some broader strokes before spending a lot of time and effort on highly detailed design work. Engaging in conversation with stakeholders at this exploratory stage creates a more inclusive process, which is far preferable to a process in which stakeholders simply grunt approval or disapproval of design deliverables.

Front-end prep chef

As we discussed earlier, front-end developers are often relegated to crude production machines that are brought into the project only after all the design decisions are made. This archaic process keeps disciplines out of sync with one another and prevents teams from working together in a meaningful way. This is a huge mistake. Including front-end development as a critical part of the design process requires changes to both project structure and team members' mentalities.

In the restaurant business, an important yet unsung role is that of the prep chef. A prep chef chops vegetables, marinades meat, and makes salads in preparation for the following day's work. By having ingredients prepared ahead of time, the kitchen staff can focus on collaboration and cooking rather than menial tasks. Without the up-front work of the prep chef, the flow of the main chefs would be interrupted and the fast pace of the kitchen would grind to a halt.

A prep chef chops vegetables, marinades meat, makes salads, and prepares other ingredients so that the main kitchen staff can focus on cooking meals and collaboration.

Front-end developers need to be the prep chefs of the web design process. **If developers aren't coding from day one of the project, there's something wrong with the process.** "But Brad," I can hear you saying, "how can I start coding if I don't know what I'm supposed to code?"

Believe me, there is plenty of front-end work to do without knowing a thing about the project's information design or aesthetic direction. In addition to setting up the development environment (such as preparing Git repositories, dev servers, CMSes, and development tools), developers can dive into code and begin marking up patterns. But what should you be marking up if you don't know anything about the design? That depends on the type of project you're working on.

Are you making an e-commerce site? You can set up site search, a shopping cart table, a placeholder product detail page, the homepage, and checkout pages. Making an online service? Start marking up the sign-up and login forms, the forgot password flow, and dashboard. And, of course, most websites will have a header, footer, and main content area. Set up shell templates and write basic markup for patterns you anticipate using. This markup will initially be crude, but it provides a crucial starting point for collaboration and iteration.

This front-end prep chef work frees up developers' time to collaborate *with* designers, rather than working *after* design is complete. With basic markup in place, developers can work with designers to help validate UX design decisions through conversations and working prototypes. They can help visual designers better understand source order and web layout, and can quickly produce a fledgling codebase that will eventually evolve into the final product.

Stop, collaborate, and listen

Let's quickly review what establishing design direction looks like across disciplines:

- **UX designers** can create lo-fi sketches to establish basic information architecture and some anticipated UI patterns.
- **Visual designers** can gather the teams' aesthetic values by conducting a 20-second gut test exercise, then create style tiles and element collages to explore initial design directions.

- **Front-end developers** can set up project dependencies, stub out basic templates, and write structural markup for patterns the team anticipates using in the project.

This work can happen concurrently but shouldn't happen in isolation. Sure, there will need to be some initial head-down time for each discipline to get set up, but all team members should be fully aware of each discipline's explorations in anticipation of working together to evolve these ideas.

> *Ideas are meant to be ugly.*
>
> - Jason Santa Maria[79]

At this early stage, it's important to stress the importance of exploration, play, and idea generation. The lo-fi nature of the techniques we just discussed help encourage this exploration, allowing team members to pursue ideas that excite them. Sometimes those ideas might be best articulated as a napkin sketch, a prototype in CodePen[80], a visual exploration in Sketch, a quick wire in Balsamiq, a motion concept in After Effects, or some combination of media and tools. **The point is for the team to generate ideas and solve problems, not to enforce a rigid order of operations.** By approaching this design exploration in a cross-disciplinary way, teams can find balance between aesthetics, technical feasibility, usability, and functionality.

Rolling up our sleeves

With a general design direction established, the team can roll up their sleeves to build the interface and its underlying design system. But how do teams turn a vague sense of direction into a beautiful, functional, usable, and complete design system?

From concept to complete

Turning explorations into finished patterns is a blurry, imperfect process. This should come as absolutely no surprise to you by this point in the book.

---

79 http://jasonsantamaria.com/articles/piles-of-ideas
80 http://codepen.io/

For the TechCrunch project, Dan Mall riffed on the team's initial design conversations to create a visual exploration for the site's header. This piece of interface was a logical place to start since the header is one of the most prominent and branded elements on the page. After a little bit of work, we hopped on a call to discuss the exploration with the client.

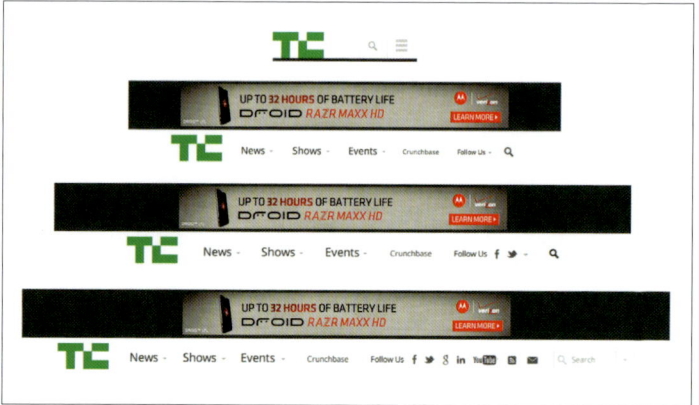

Dan Mall created an element collage to explore an aesthetic direction for the global header.

Even though this design artifact was a simple in-progress exploration, we were able to have important conversations about the header's aesthetics, hierarchy, and suggested functionality. Because the header was presented sans context, we were able to discuss the issues pertaining to the header without stakeholders' focus wandering to other page elements.

Though the client didn't know it, I had been building out a working HTML version of the header behind the scenes in Pattern Lab.

Using Dan's exploration as a reference, I created an HTML version of the global header in Pattern Lab. This grayscale prototype helped us demonstrate interactivity and how the header would adapt across the resolution spectrum.

This grayscale prototype allowed us to demonstrate interactivity and responsiveness, which led to even more discussion. Collectively we proposed changes to the header's layout and functionality, and I was able to make changes using the browser's development tools during the call. Suddenly, the entire team and stakeholders were actively participating in the design process!

With input from the stakeholders and team, we iterated over the header pattern to massage the layout, IA, aesthetic details, and functionality to arrive at the solution we ultimately launched with.

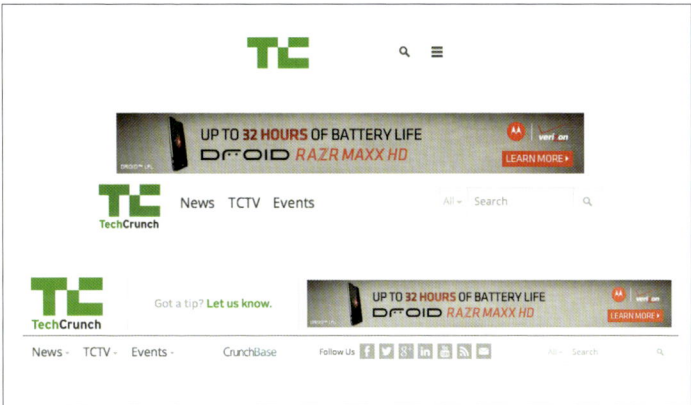

The header we launched with was the culmination of plenty of conversations and decisions around the pattern's content, design, and functionality.

Obviously the header pattern doesn't exist in a vacuum. Within Pattern Lab, the header was included in every template using Mustache's include pattern that we discussed in chapter 3.

```
{{> organisms-header }}
```

This allowed us to view the header within the context of the rest of the pages, sketchy as they initially were. So while we were focusing on designing one specific pattern, we were simultaneously taking into account the context of where that pattern would be employed.

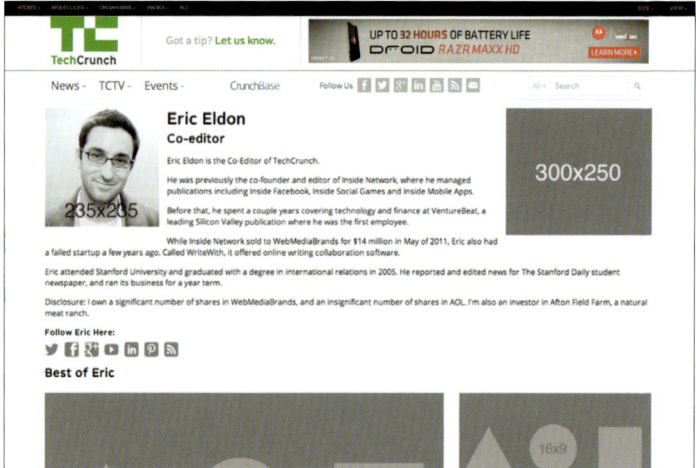

In a more iterative process, there will be instances where some patterns are further developed than others. Seeing a partially done page might look unusual out of context, but communication between the team and stakeholders should alleviate confusion.

Initially, in-browser designs tend to look crude at best, which is A-OK. The intention is to stub out the template's basic information architecture in the browser, define patterns, wire up those patterns using includes, and begin the patterns' general markup. With that work in place, the team can collectively begin styling specific patterns and refining the overall structure.

Seeing these partially designed prototypes might look unusual to those used to more traditional, pixel-perfect design deliverables. But it's far more important to communicate progress than a false sense of perfection, which is why rolling updates are preferable to big reveals.

The role of comps in a post-PSD era

Up until this point we've been talking about establishing a general aesthetic direction and then designing some patterns to experiment with the application of that aesthetic direction. These relatively lo-fi tactics allow teams to explore freely, iterate quickly, and get feedback sooner.

But I'll never forget this client feedback we received on the first pattern-driven project I worked on: "These element collages look great, but it's like you're asking me to comment on how beautiful a face is by showing me the nose."

If you've gotten to this point in your process, congratulations! Feedback like this means they're salivating for more, so now that you've captured a general aesthetic direction you can safely put those explorations into context. That likely involves creating full static comps.

Listen to the chatter around "designing in the browser" and you'll undoubtedly hear that *Photoshop comps are the devil incarnate*. Which, of course, isn't true. Throughout this book we've discussed the importance of breaking things down into their atomic elements while simultaneously building up a cohesive whole. **Static comps are effective at painting a full picture of what the UI could look like.** The trick is knowing *when* to paint those full pictures, and knowing how long to dwell in static design documents.

Dan Mall created a full comp to demonstrate what a featured article template for TechCrunch might look like. This artifact was used to show the design system in context and get approval for the overarching design. Subsequent design revisions would be handled in the browser.

For the TechCrunch project, we created a comp for the article template only *after* the client was feeling good about our element collage explorations. Creating full comps requires a lot of effort, which is why we established the design direction first to mitigate the risk of all that full-comp effort going straight into the trash if we got it totally wrong.

Comps, like any other design artifact, are used to facilitate a conversation with the project stakeholders. If their feedback is, "This feels all wrong," then it's back to the drawing board to create a new comp. But if their feedback suggests, "Can we move this from here to here? Can we add a gray border around the article text? Can we increase the size of this image?" that's a sign the overall direction is in good shape and those relatively minor issues can be addressed in the browser.

In-browser iteration

Static comps can be great for shaping the overall aesthetic direction of a template, but users will ultimately view and interact with the experience in a browser. That's why designs should be quickly translated into the final environment and iterated on there.

Working in the browser allows teams to address layout issues across the entire resolution spectrum, design around dynamic data (such as variable character lengths, image sizes, and other dynamic content), demonstrate interaction and animation, gauge performance, factor in ergonomics, and confront technical considerations (such as pixel density, text rendering, scrolling performance, and browser quirks). Static design comps cannot deal with all these considerations, so they should be treated merely as hypotheses rather than set-in-stone specifications. Only when transferred to the browser can any design hypothesis truly be confirmed or rejected.

> *Let's change the phrase "designing in the browser" to "deciding in the browser."*
>
> – Dan Mall[81]

---

81 https://the-pastry-box-project.net/dan-mall/2012-september-12

Once the designs are in the browser, they should stay in the browser. At this stage in the process, the point of production shifts to team members adept at crafting HTML, CSS, and presentational JavaScript. Patterns should be created, styled, and plugged in wherever they're needed. Designers can react to these in-browser implementations and can create spot comps in static tools to help iron out responsive wrinkles at the organism level. This back-and-forth between static and in-browser tools establishes a healthy loop between design and development, where the front-end code becomes more solid and stable with each iterative loop.

This illustration by Trent Walton of Paravel perfectly articulates a more iterative design and development process. By getting designs into the browser sooner, teams can iterate over the design and address the many considerations that can only be dealt with once the design is in the browser.

The beautiful thing about a pattern-based workflow is that as each pattern becomes more fully baked, any template that includes the pattern will become more fully baked as well. That means the level of effort to create new templates decreases dramatically over the course of the project, until eventually creating a new template mostly involves stitching together existing patterns.

Bring it on home

The design system is taking shape and the team is cooking with gas to bring the project home. At this stage, UI patterns are well established, the team is taking some final steps to tighten everything up and prepare for launch.

UX designers are hitting the prototype hard to make sure the flows and interactions are all logical and intuitive. Visual designers are combing over the interface and proposing design tweaks to the UI to polish up the design. Front-end developers are testing the experience in a slew of browsers and devices while also addressing design feedback. Back-end developers are hard at work integrating the front-end UI into the CMS (we'll talk more about the relationship between front-end and back-end in chapter 5). And, of course, the clients and stakeholders are making last-minute demands – I mean suggestions – about the design and content. The whole team is contributing documentation for the style guide, cleaning up the patterns in the pattern library, and working hard to get the website off the ground.

Then – seemingly in the blink of an eye – the website and accompanying design system launch. Champagne is poured, high-fives are exchanged and, of course, post-launch bugs are squashed. Users visit the new site to find a beautiful, functional, consistent, and cohesive experience that undoubtedly makes them weep tears of joy. Mission accomplished.

What began as a giant slab of rock is now a finely polished sculpture, thanks to a ton of hard work, genuine collaboration, constant communication, and plenty of iteration. Moreover, in addition to a brand-spanking-new website, the team leaves behind a flexible, deliberate UI system bundled up in a beautiful style guide.

This chapter explored everything that goes into making an effective UI design system. In the next chapter, we'll discuss how to make sure that design system continues to be successful in the long run.

Chapter 5
Maintaining Design Systems

Making design systems stand the test of time

And they made a design system, delivered a style guide, and lived happily ever after. Right?

Not quite.

There's a very real risk that a style guide will end up in the trash can right alongside all the PSDs, PDFs and those other static artifacts of the design process. Despite everyone's best intentions, all the time and effort that went into making a thoughtful design system and style guide can go straight down the drain.

How can that be?

> *A style guide is an artifact of design process. A design system is a living, funded product with a roadmap & backlog, serving an ecosystem.*
>
> - Nathan Curtis[82]

An artifact is something you'd find on an archaeological dig or in a museum, whereas a system is a living, breathing entity. A style

---

82 https://twitter.com/nathanacurtis/status/656829204235972608

guide can provide documentation and serve as a helpful resource, but the simple existence of a style guide doesn't guarantee long-term success for the underlying design system. A design system needs ongoing maintenance, support, and tender loving care for it to truly thrive.

Changing minds, once again

We've already discussed the importance of resetting everyone's expectations to establish a more collaborative, pattern-driven workflow. To save our style guides from the bowels of a trash can, we must once again fundamentally rewire people's brains.

What is it we're making again?

We *think* we merely design and build websites and apps. And that's true for the most part. After all, that's what our clients pay us to do, and the products we create are the vehicles that generate money and success for our organizations. It seems natural to focus on the final implementations rather than the underlying system. The live products remain the primary focus of everyone's attention, while any pattern library exists as an offshoot that simply provides helpful documentation.

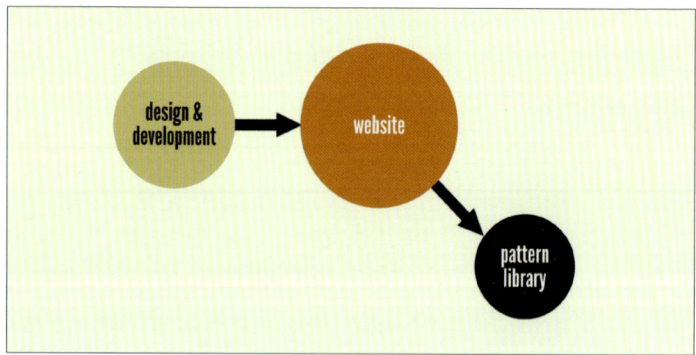

The problem with this mindset is that you can almost see that pattern library snapping off and sliding into the abyss. **Once the pattern library ceases to reflect the current state of the products**

it serves, it becomes obsolete. And when the pattern library managing the design system is no longer accurate, the website maintenance process devolves into a smattering of hotfixes and ad hoc changes, ruining all the thoughtfulness that went into creating the original design system.

To set ourselves up for long-term success, we must fundamentally shift our outlook around what we're actually creating. Rather than thinking of final applications as our sole responsibility, we must recognize that the design system is what underpins our final products *and* pattern libraries.

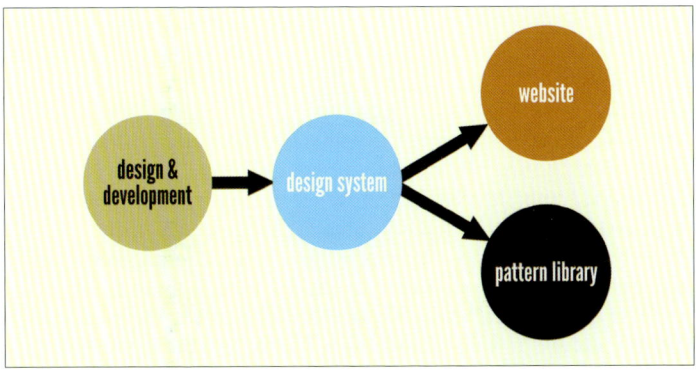

This "design system first" mentality inserts a bit of friction into the maintenance process, and that **friction can be friendly**. It forces us to step back and consider how any improvements, client requests, feature additions, and iterations affect the overall system rather than only a sliver of the whole ecosystem.

Say you're working on an e-commerce site, and you run a test that finds a custom-styled dropdown menu on the product detail page isn't performing as well as the browser's default dropdown menu. One course of action is to simply remove the custom-styled dropdown from that particular page and call it a day. However, considering the entire design system rather than just the product detail page might cause you to take a step back and wonder, "If this custom dropdown menu isn't performing well here, perhaps it's not performing well elsewhere." After digging into the issue further, you find the best course of action is to globally modify the dropdown pattern in the design system to remove the custom styling. Now, anywhere the dropdown pattern appears will reflect those changes and will likely see similar performance improvements.

That is just one example of how design-system thinking can lead to broader, more considered changes. **Broken behavior and opportunities to enhance the UI will often be realized at the *application* level, but those changes should often be acted on at the *system* level.** Adding this bit of friendly friction into your workflow ensures improvements are shared across the entire ecosystem, and prevents the system from being eroded by a series of one-off changes.

Done and done

Another expectation we must revisit is our definition of *done*. Creating things for print and other physical media involves making permanent, tangible objects. That sense of finality simply doesn't exist in the digital world, which means change can happen with much less effort and friction than other media. **Clients, colleagues, and stakeholders should embrace the pliable nature of the digital world to create living design systems that adapt to the ever-shifting nature of the medium, user needs, and the needs of the business.**

This shift in thinking fundamentally affects the scope of our work. Folks working in the client services business are often used to delivering a project in a tidy package then riding off into the sunset. Internal teams don't fair much better, since they tend to float from one initiative to the next. Whether you're part of an internal team or you're an external gun for hire, I'm guessing you've experienced the shortcomings of project-based work. We tend to talk about a future that never comes, and instead we set it, forget it, then move on to the next shiny project.

If we're committed to creating genuinely useful work that truly meets the needs of our clients and organizations, we must fundamentally redefine the scope of our work. As Nathan Curtis says, a design system shouldn't be a *project* with a finite scope, but rather a *product* meant to grow and evolve over time:

> *Focusing on style guide delivery as the climax is the wrong story to tell. A system isn't a project with an end, it's the origin story of a living and evolving product that'll serve other products.*
>
> – Nathan Curtis[83]

The web is never done, and the creation of a design system is merely the first step in a long (and hopefully fruitful!) journey. A design system should be a long-term commitment with the ambitious goal of revolutionizing how your organization creates digital work. Exciting, eh?! So how do we make sure that happens?

Creating maintainable design systems

As you embark on this pattern-paved journey, let's talk about things you can do to craft a design system that sets up your organization for long-term success. How do you create a design system that takes root and becomes an essential part of your organization's workflow? What pitfalls do you need to be on the lookout for? How do you ensure the design system yields big results? To set up your design system for long-term success, you need to:

- Make it official.
- Make it adaptable.
- Make it maintainable.
- Make it cross-disciplinary.
- Make it approachable.
- Make it visible
- Make it bigger.
- Make it context-agnostic.
- Make it contextual.
- Make it last.

Let's dive into each one of these points in a bit more detail.

83 https://medium.com/eightshapes-llc/a-design-system-isn-t-a-project-it-s-a-product-serving-products-74dcfffef935#.4umtnfxsx

Make it official

Your initial style guide may begin its life as a side project, the result of a weekend hackathon, or as the brainchild of one or two ambitious team members. As we discussed in the previous chapter, your client or boss doesn't even have to know that you're creating a thoughtful design system and accompanying pattern library. Remember: ask forgiveness, not permission!

Organic beginnings are all well and good, but in order to establish a truly impactful design system that creates long-term success for your organization, **the design system needs to evolve into an officially sanctioned endeavor**. That means thinking of it as a true product rather than a simple side project, and consequently allocating real time, budget, and people to it.

Convincing stakeholders to commit large chunks of money, time, and resources up front for a design system can be extremely challenging. So what are we to do? Here's my advice:

1. Make a thing.
2. Show that it's useful.
3. Make it official.

Let's break down these steps a bit further.

1: Make a thing

You have to start somewhere, and having something started is better than nothing at all. Pick a project that would be a great pilot for establishing your design system; follow a process similar to the one discussed in chapter 4; think about the atomic design mental model detailed in chapter 2; and you'll end up with a solid foundation for a thoughtful design system and pattern library that helps your team work more effectively.

Take the time to package your UI patterns in a pattern library and get it ready to be shopped around. I've talked to several ambitious team members who have established the basic gist of their pattern library over the course of a weekend. This effort makes all the difference in the world since it provides something tangible for stakeholders to react to. Again: *show, don't tell.*

2: Show that it's useful

With a nascent-yet-tangible design system in place, you can have more meaningful conversations with the people who control money, scheduling, and resources. You can discuss exactly how the design system helped save time and money (see "Pitching Patterns" in chapter 4), then paint a picture of how those benefits would scale even further if the organization invested in an official, full-fledged design system.

Get team members from different disciplines to back you up and discuss the initial success of the system, and also pull in others who are sympathetic to the cause who would stand to benefit from an expanded design system.

3: Make it official

You've proved the value of your initial design system and presented a roadmap for how to make it even better. With any luck your organization will commit to making the design system an Official Thing.

With approval from the highest levels, you're now able to put a plan into action that involves: allocating or hiring people to work on the design system; developing a plan to make it more robust; establishing a clear governance strategy; and laying out a product roadmap.

It's worth pointing out that things may not shake out the way you hoped. Despite demonstrating real value and presenting a concrete plan of action, higher-ups still might shoot your initiative down. **Don't be discouraged.** You may have lost the battle, but you can certainly win the war. Your team should continue to grow and extend the design system in whatever capacity you can until its value becomes undeniable. As more people benefit from the system, you'll end up with a grassroots-supported system that can help push the endeavor through.

Establishing a design system team

With the design system initiative approved, it's now time to put the right people and processes in place to ensure the system flourishes for your organization.

Design system makers and users

First things first. It's important to recognize that **there will inevitably be people at the organization who help** *make and maintain* **the design system, and there will be people who will be** *users* **of the design system.** There may be overlap between these two groups, but establishing the roles of makers and users is important nonetheless.

When I talk about establishing a more collaborative process like the one I detailed in the previous chapter, I inevitably hear people who work at large organizations say, "But Brad, we have hundreds (or even thousands) of developers working on our products. Getting all those people to collaborate and contribute like that would be far too difficult."

They're likely right. It would be ideal if the entire organization adopted nimbler, more collaborative processes, but the daunting logistics around such an effort makes it improbable. But here's the thing: not *everyone* in the organization needs to contribute directly to the design system, but someone (or more likely, some people) must take ownership of it.

The design system *makers* **are the ones who create, maintain, and govern the system**, and they need to work closely together to ensure that the system is smart, flexible, scalable, and addresses the needs of the users and business. **The design system** *users* **are the teams across the organization who will take the system and employ its interface patterns to specific applications.**

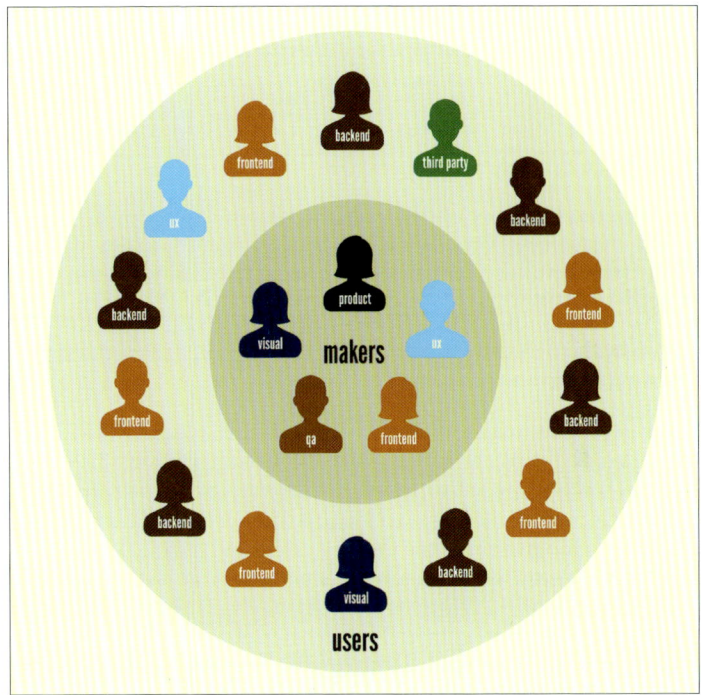

Design system makers and users.

The design system *makers* and design system *users* need to maintain a close working relationship to ensure the patterns defined within the system serve the needs of the applications, and that all documentation is clear. **Makers provide a birds-eye perspective of the entire ecosystem the design system serves, while users provide an on-the-ground perspective focused on specific applications of the system.** Jina Bolton of Salesforce sums up the relationship between makers and users quite nicely:

> *The Design System informs our Product Design. Our Product Design informs the Design System.*
>
> - Jina Bolton, Salesforce[84]

84 https://medium.com/salesforce-ux/the-salesforce-team-model-for-scaling-a-design-system-d89c2a2d404b

Both outlooks are critical to the success of the design system, which is why it's so important for makers and users to have a healthy relationship that involves frequent communication and collaboration.

Design system makers

Who updates the design system? Who approves changes? Who communicates with the users of the design system to make sure it's addressing their needs? Who gets to decide which patterns stay, go, or need tweaking?

The answers to these questions will very much depend on the size and setup of your organization.

Large organizations are able to dedicate serious resources to managing design systems. Salesforce, for example, maintains an official *design systems team*, which currently includes about a dozen full-time employees, last I heard. That dedicated team is responsible for governing the design system and making sure it's meeting the needs of the internal product teams, as well as external developers who build things on the company's platform. When a design system is serving literally thousands of users, it's a smart idea to dedicate at least a few full-time employees to manage and expand the system.

Smaller organizations most likely don't have the luxury of building an entire team to service a design system. Team members in smaller organizations have to wear many (hopefully stylish!) hats out of necessity, so governing the design system will likely become another responsibility. This may sound like an added burden ("Oh great, yet another thing I'm responsible for that doesn't involve a pay raise!"), but this particular hat should be a joy to wear as it improves the efficiency and quality of all other work. Hooray for design systems!

Typically, design system makers at smaller organizations will be senior-level staff who have the experience to make thoughtful decisions, and the authority to enforce the design system.

And then there are **external agencies, contractors, and consultants**. What is the role of a third party when it comes to long-term maintenance of a client's design system? On one hand, external partners are at a bit of a disadvantage since they don't actually work for their client's organization. A successful design system needs to become part of an organization's DNA, and since third parties exist outside the company's walls, their influence is intrinsically limited.

But on the other hand, **external parties can often provide a sense of perspective** that's hard to see while working inside a company. This is where outsiders can really shine. In my work as a consultant, I work with organizations to establish long-term design system maintenance strategies, and help get the right people and processes in place. While the long-term success of the system will ultimately be up to the organization, third parties can teach them to fish and provide important strategic guidance, feedback, and perspective.

Design system users

Who are the people responsible for using the design system to build new features and applications? Who are the people who talk with the system makers to report issues and request features?

Once again, the answers to these questions will largely depend on your organization's size and structure.

Design system users may be the same team creating the design system, separate development teams within your organization, junior-level designers and developers, partner agencies, external development shops, or other third-party teams.

Users' proximity to and involvement in the creation of the design system will undoubtedly vary. You may work on a singular product at a scrappy startup, so your small team could be simultaneously creating and using the design system. Or you may work at a large multinational corporation with development teams and third-party partners scattered all across the globe. If this is the case, design system makers and users may seldom (or ever) meet, which means that helpful documentation and a sharp birds-eye perspective become that much more important.

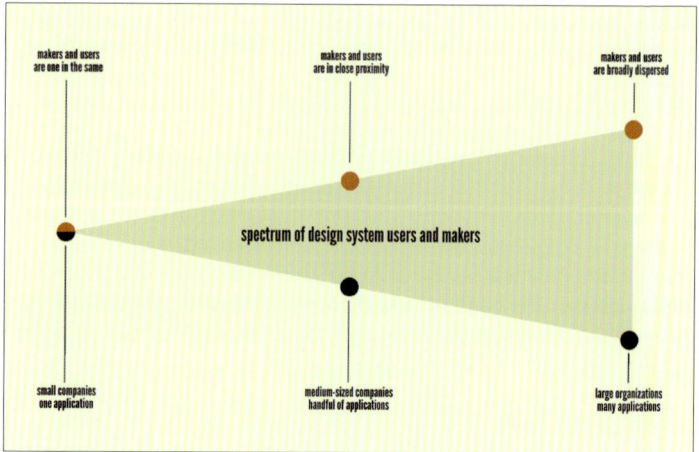

There is a spectrum of potential relationships between design system users and makers, and the size and makeup of your company will undoubtedly shape those relationships.

One of the biggest advantages of establishing a thoughtful design system is that it allows organizations to scale best practices. If all those best practices – responsiveness, accessibility, performance, UX, ergonomics, and so on – are baked into the system, users can simply plug in the patterns and reap the rewards. This means design system users don't have to be senior-level designers or developers to produce good work; the design system serves as a quality control vehicle that helps users apply best practices regardless of each individual's skill level.

Design system team makeup

A cross-disciplinary team should be established to properly manage, maintain, and extend the system. All disciplines at an organization – UX designers, visual designers, content strategists, front-end developers, back-end developers, product managers, project managers, executives, and other stakeholders – have unique perspectives that can undoubtedly inform and shape the work. Incorporating these perspectives into the design system is important, but doesn't necessarily require every discipline to be constantly involved in developing it.

There will inevitably be disciplines that actively do the work, while others may take on more of an advisory role. **Those responsible for designing and building the user interface – UX designers, visual designers, front-end developers – will likely serve as the hands that do the work and make updates to the design system.** They should work collaboratively (as detailed in chapter 4) and coordinate with other disciplines to ensure that the system reflects the values and considerations of the entire business.

Other people may not be the ones actively doing the work, but must be consulted to make sure their perspectives are properly reflected in the system. Back-end engineers need to make the team aware of any architectural decisions that would affect the front-end UI; executives need to make the team aware of important initiatives that will affect the role and utility of the system; and, of course, design system users need to coordinate with the makers to ensure the system serves the needs of individual applications.

Make it adaptable

Change is the only constant, as they say. **The *living* part of a living design system means that it needs to roll with the punches, adapt to feedback, be iterated on, and evolve alongside the products it serves.**

A misconception about design systems is that once they're established, they become an omnipotent and unchangeable source of truth. Thinking in such a rigid way is a surefire way to have your design system effort backfire. If users feel handcuffed and pigeonholed into using patterns that don't solve their problems, they'll perceive the design system as a unhelpful tool and start searching elsewhere for something that will better address their needs.

Creating a clear governance plan is essential for making sure your design system can adapt and thrive as time goes on. A solid governance strategy starts by answering some important questions about handling change. Consider the following:

- What happens when an existing pattern doesn't quite work for a specific application? Does the pattern get modified? Do you recommend using a different pattern? Does a new pattern need creating?
- How are new pattern requests handled?

- How are old patterns retired?
- What happens when bugs are found?
- Who approves changes to the design system?
- Who is responsible for keeping documentation up to date?
- Who actually makes changes to the system's UI patterns?
- How are design system changes deployed to live applications?
- How will people find out about changes?

There are likely many more specific questions to answer, but the point is your team should have answers and processes in place to address inevitable changes to the system.

As mentioned a few times already, frequent communication and collaboration between makers and users is key for successfully governing your design system. **Make it as easy as possible for users and makers to communicate.** Set up a design system Slack or Yammer channel, establish regular office hours, make sure your bug ticket software helps facilitate conversation, and keep the doors open for ad hoc chats and calls. If users are stuck on something, they should know exactly where and who to turn to for help.

In addition to informal day-to-day conversation between makers and users, **schedule regular "state of the union" meetings to review the design system** with makers, users, and other key stakeholders. Discuss what's working, be honest with what needs to be improved, and review priorities and the roadmap to make sure the system is serving the needs of the business. These regular checkups are especially helpful for keeping stakeholders up to speed, since they often aren't involved in the day-to-day of the design system's operations.

Making changes to patterns

A critical part of design system maintenance is ensuring that UI patterns stay up to date, embrace evolving design and development best practices, and continue to address the real needs of the organization.

Developing a strategy for handling pattern changes is crucial, which is why Inayaili de León Persson and the Canonical web team spent time to map out their strategy as they created the Vanilla front-end framework.

> We thought that it would be good to document the process that a pattern should follow in order to become a Vanilla pattern, so after a little bit of brainstorming, we created a diagram that shows the different steps that should be taken from before submitting a pattern proposal to its full acceptance as a Vanilla pattern.
>
> – Inayaili de León Persson, Canonical[85]

The result is a gorgeous decision tree that maps out exactly what processes need to happen to add a new pattern to the design system.

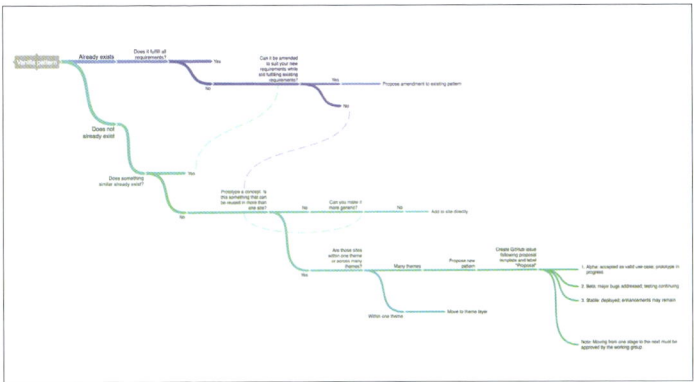

The Canonical web team mapped out the decision process used to manage updates and additions to patterns in the Vanilla front-end framework.

The three types of change that can happen to patterns in a design system are modification, addition, and removal.

Modifying patterns

UI patterns can and should be modified for a number of reasons: feature additions, bug fixes, subtle or major visual design tweaks, performance improvements, accessibility enhancements, code refactoring, UX best practice updates, and so on.

85 http://design.canonical.com/2016/07/getting-vanilla-ready-for-v1-the-roadmap/

The design system maintainers need to understand why and when to tweak patterns, how to go about making those changes, and how to roll out those improvements into individual applications.

Keeping patterns fresh is essential for the long-term health of the design system. Nobody wants to use and maintain a Web 2.0-looking design system full of bevels and crusty code!

Adding patterns

As smart as your team surely is, it's quite possible you won't think of every conceivable pattern to include in your design system right out of the gate. As the system is applied to more products, gaps will inevitably emerge where the needs of the application aren't solved by existing patterns. In such cases, it will become clear that new patterns will need created to address these needs.

Care should be taken when adding patterns to the library. **If every whim results in a brand new pattern, the design system will become a bloated and unwieldy Wild West.** It's worth asking if this is a one-off situation or something that can be leveraged in other applications.

Perhaps you may want to assume a one-off until a different team encounters a similar use case. If the team working on Application 2 looks at Application 1 and says, "I want that!" perhaps that's a good indicator that a one-off pattern should be added to the pattern library.

Removing patterns

Patterns can be deprecated for a number of reasons. Perhaps you discover through use that a particular pattern is a terrible idea. Hindsight is 20/20, my friend. Maybe the industry has moved away from a pattern for UX or technical reasons. Perhaps a pattern sat there unused by any application for ages. Maybe users reported back with a lot of negative feedback about working with a particular pattern.

Having a plan for deprecating patterns is a great idea. But how do you remove patterns from the design system without pulling the rug out from under people relying on those patterns in their applications? To address this issue, Salesforce created a neat utility called Sass Deprecate[86] that flags patterns that are heading to the chopping block in the near future. Through some clever use of Sass variable flags and styling, the maker team can give a heads-up to users that a particular pattern is being deprecated, and recommend an alternative pattern instead.

Make it maintainable

With all this talk about modifying, adding, and removing patterns, you may be wondering, "How the hell are our applications supposed to actually keep up with all these changes?!" And in asking that question, you will have stumbled on to one of the biggest challenges organizations face in successfully maintaining a design system.

> *The biggest existential threat to any system is neglect.*
>
> – Alex Schleifer, Airbnb[87]

Many systems fall into a state of disrepair because the effort required to make updates is far too high. If it's difficult and time-consuming to update patterns, documentation, and applications, people will eventually get so frustrated that they stop making the effort and the design system will begin its drift into oblivion. **Making updates to UI patterns, documentation, and applications should be as frictionless as possible,** so reducing this friction should become a high priority for the design system team. This involves careful consideration from both technological and workflow standpoints.

In search of the holy grail

The design system holy grail involves creating an environment where the pattern library and live applications are perfectly in sync. The idea is that you should be able to make a change to a UI pattern and see that change automatically reflected in both the pattern library *and* anywhere the pattern is included in production.

---

86 https://github.com/salesforce-ux/sass-deprecate
87 http://airbnb.design/the-way-we-build/

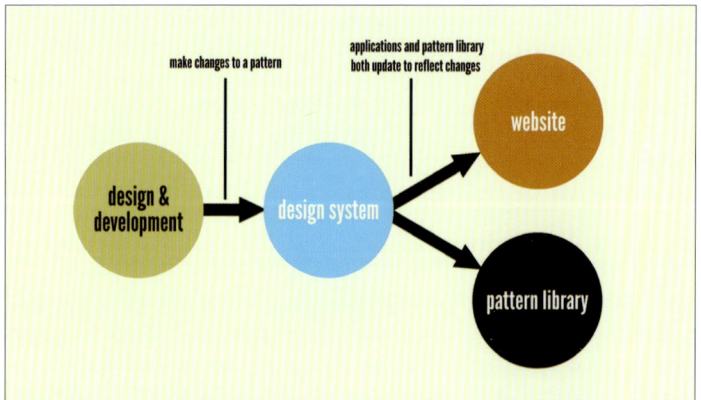

The holy grail of design systems is an environment where making changes to a UI pattern updates both the pattern library and production applications simultaneously.

This technique removes any duplication of effort and ensures the pattern library and the applications using the patterns remain synchronized. Sounds like a dream, right?

As it turns out, this dream can be a reality. Lonely Planet, the travel guide company, was one of the first to establish a holy grail design system called Rizzo[88]. Through some smart architecture, they created an API for their UI patterns that feeds into their production environment as well as their pattern library. The result is a centralized design system that ensures their live application and documentation remain perfectly in sync.

This approach is no easy task, as it requires sophisticated technical architecture, smart people to set it all up, and a relatively centralized organizational culture. How you go about chasing the holy grail – or even if you can achieve it – is dependent on a whole load of factors, including your technical architecture and organizational makeup.

88 http://rizzo.lonelyplanet.com/

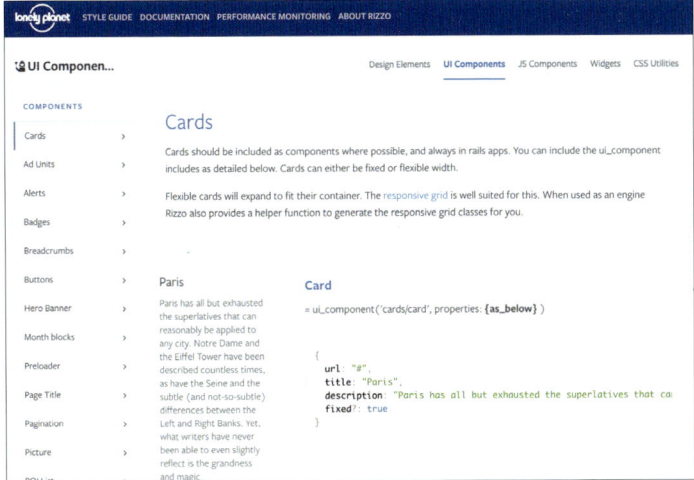

Lonely Planet created an API for its UI patterns that is consumed by both their pattern library and production environment. By constructing their design system in this manner, changes to UI patterns are automatically reflected in both the pattern library and production environment.

Clearing technical hurdles

Keeping a pattern library in sync with production environments requires sharing code in a smart, scalable, and maintainable way. Detailing all the different technical strategies and considerations around the holy grail would necessitate its own book, but let's cover a few important areas around keeping front-end code in sync.

The front-end of things

A UI design system manifests itself as the front-end of a web experience, which is comprised of HTML, CSS, and JavaScript. How we get that front-end code into a production environment, with complex application logic and back-end code, is the task at hand.

In his article "Chasing the Holy Grail[89]," web developer Marcelo Somers details various technical approaches to achieving the holy grail. He highlights the pros and cons of each strategy for feeding a

89 https://medium.com/@marcelosomers/chasing-the-holy-grail-bbc0b7cce365#.ay1xeej7d

design system into applications to keep both codebases in lockstep. While I won't detail each of Marcelo's strategies, it's worth noting there is a spectrum of approaches to choose from: crude, manual front-end code copying-and-pasting on one end, to baking the pattern library directly into the production environment on the other.

In my experience, I've found that sharing CSS and presentational JavaScript with production environments is relatively easy, while sharing markup is tough. Because CSS and JavaScript tend to get compiled into a single file (or perhaps a handful of files), it becomes possible to throw them onto a CDN and then simply link to those files in each application. Marcelo explains how to do this while keeping versioning in mind:

> You'd provide development teams with a versioned URL (e.g., http://mycdn.com/1.3.5/styles.css) and upgrading is as simple as bumping the version number in the URL.
>
> - Marcelo Somers[90]

Sharing CSS and JavaScript is all well and good, but where things get tricky is when you want to share markup between environments. Why? you ask. Well, markup and back-end logic are often intertwined in an application's codebase, which tends to make it difficult to simply copy and paste markup between your pattern library and production environments. Thankfully, there are ways around this problem.

Bridging the markup gap with templating languages

Using HTML templating languages (such as Mustache, Handlebars, Twig, Underscore, Jade, Nunjucks, and a slew of others) makes markup more portable and dynamic. Templating languages separate structure and data, and supercharge our HTML to keep us from having to write the same markup patterns over and over again. The good news is that many CMSes and application environments also make use of templating languages to serve up front-end markup.

---

90 https://medium.com/@marcelosomers/chasing-the-holy-grail-bbc0b7cce365#.ay1xeej7d

The templating language can serve as the bridge between your pattern library and production environments. If you use a templating language to create the patterns in your design system (something we discussed at length in chapter 3), you can easily share those patterns with production environments that utilize the same templating engine.

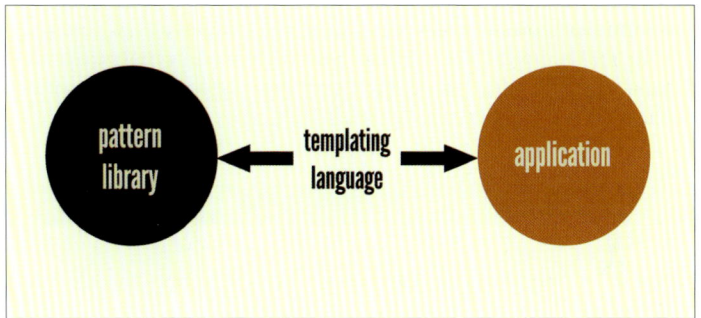

A templating language like Mustache, Handlebars, Underscore, Jade, and others can serve as a bridge that allows front-end code to be shared between the pattern library and production application.

The team at Phase2 Technology achieved the holy grail by using Pattern Lab as their pattern library development tool and Drupal[91] as their content management system. Because both Pattern Lab and Drupal support the popular Twig[92] templating engine, Phase2 is able to easily share patterns between the two environments, ensuring their clients' pattern libraries and production builds are always in step with each other.

> By using the same templating engine, along with the help of the Component Libraries Drupal Module, the tool gives Drupal the ability to directly include, extend, and embed the Twig templates that Pattern Lab uses for its components without any template duplication at all!
>
> – Evan Lovely, Phase2 Technology[93]

91 https://www.drupal.org/
92 http://twig.sensiolabs.org/
93 https://www.phase2technology.com/blog/introducing-pattern-lab-starter-8/

Is your culture holy grail compatible?

You may have read that last section and thought, "That's amazing! My company needs this now!" While holy grail systems are indeed great, there are reasons why you may not be able to automagically keep your production environments and pattern library in sync. Perhaps your organization creates tons of digital products on many different platforms using wildly different technologies. Maybe you're a giant multinational company scattered all over the world. Maybe your company has an extremely decentralized, autonomous culture. Or maybe you're a gigantic federal government.

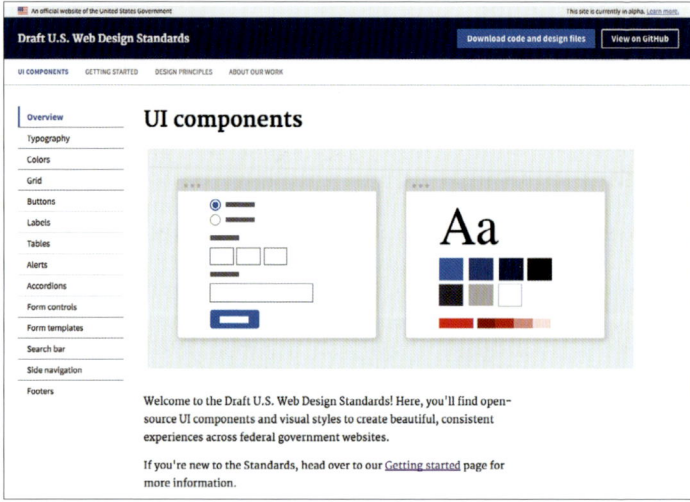

The Draft U.S. Web Design Standards are the design system for the United States federal government.

The U.S. government's design system – called the <u>Draft U.S. Web Digital Standards</u>[94] – is a collection of UI components and visual styles created to help people making government websites build more consistent UIs. The design system provides markup and styles for users to download and weave into their applications. It would certainly be amazing to see a holy grail design system implemented at such a gigantic scale, but as you might imagine, that's a pretty tall order. The vastness and decentralized nature of the organization

---

94 https://standards.usa.gov/

means that a synchronized pattern library isn't really achievable without some dramatic restructuring of how federal government websites get built.

If a relatively scattered, decentralized culture is your reality, don't be disheartened! Even getting *some* design system in place – a handful of go-to UI patterns, some helpful documentation, and guiding principles – can show your organization the light that points towards the grail. As we've discussed throughout this chapter, these efforts should be ongoing, and before you can run you must first learn to crawl.

Make it cross-disciplinary

Style guides often jump straight into code snippets and pattern usage for the benefit of the design system users. Of course, a pattern library needs to be helpful for the people actually making use of the patterns, but **treating a style guide solely as a developer resource limits its potential.**

A style guide has the opportunity to serve as a watering hole for the entire organization, helping establish a common vocabulary for every discipline invested in the success of the company's digital products. Establishing this common vocabulary can lead to more efficient work, better communication, and more collaboration between disciplines across the organization. That's why the style guide should be an inviting place for everybody, not just design system users.

Take the carousel (please!). This component is amazingly complex from an organizational standpoint. A homepage carousel on an e-commerce website requires input from a host of disciplines across the organization. Business owners and editorial staff must choose products to be featured in the carousel. Copywriters must ensure the copy is effective and stays within the constraints of the design. Art directors need to make certain the aesthetic design is pleasing and the product photography is legible across every screen size. UX designers have to confirm the functionality and controls are intuitive. Front-end people must be sure the component is responsive, accessible, and performant. Back-end developers need to ensure the component is properly wired up to the back-end system. You get the idea.

A homepage carousel on a site like Walmart requires input from many different disciplines and stakeholders. A style guide can help gather those different perspectives under one roof.

A well-crafted style guide can help manage all these moving parts and ensure the many perspectives that influence each pattern are properly documented in the style guide. Make the pattern library accessible to every discipline, and think about how to make it easy and inviting for different disciplines to contribute to the documentation.

Make it approachable

It should come as a surprise to no one that people tend to gravitate towards attractive things. A big part of making a style guide a cross-disciplinary resource is ensuring the container that houses your pattern library and other documentation is good-looking, inviting, and easy to navigate.

Taking the time to craft an attractive home for your style guide and documentation can lead to more usage, help build awareness, help create organizational investment, and help get non-developers' eyeballs on the style guide. All of this contributes to that important shared vocabulary that leads to better cross-disciplinary collaboration.

But creating a great-looking, intuitive style guide experience doesn't just happen, and this can be problematic when getting

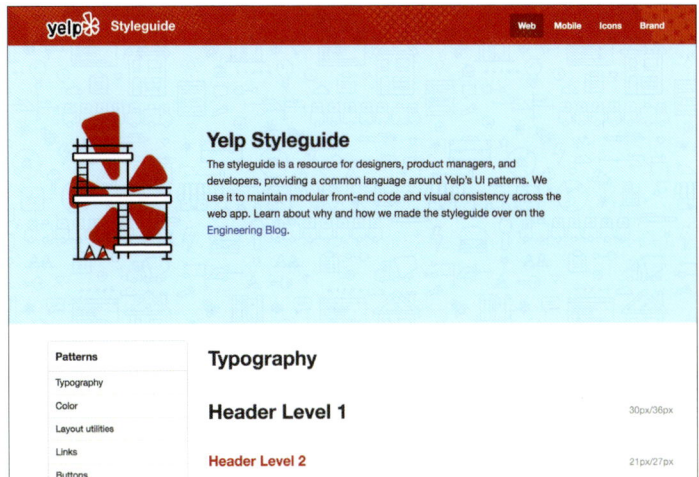

Yelp's style guide has an attractive, friendly front page that explains what the resource is, who it's for, and how to use it.

a style guide off the ground. If teams think that making a useful style guide involves making some Big, Official Thing with custom branding and a glossy website, they might be deterred from ever starting the initiative. So remember:

1. Make a thing.
2. Show that it's useful.
3. Make it official.

Creating a useful design system should be the team's first priority. Building a happy home to contain it all might not happen straightaway, but should become a bigger priority once the design system becomes official. Making a good-looking style guide isn't just design for design's sake; **it reflects an organization's commitment to making and maintaining a thoughtful, deliberate design system.**

Make it visible

Visibility is critically important to the ongoing health of your design system. Such an important endeavor shouldn't be tucked away in a dark corner of your intranet. What steps can you take to ensure the design system remains a cornerstone of your design and development workflows?

Design system evangelism

You can create the best style guide in world, use the most sophisticated technology, have an amazing team in place, and have excited users, but if you don't actively promote the design system and communicate changes, the entire effort will suffer greatly.

Evangelizing your design system efforts can and should happen even before the system is off the ground. At the onset of your project, you can set up places to document progress of the project to help garner awareness and excitement for the design system effort. One client of mine set up an internal blog to publish updates to the project, as well as a design system Yammer channel where developers and other interested parties can share ideas, address concerns, give feedback, and ask questions. Establishing a culture of communication early in the process will increase the likelihood of the design system taking root.

Communicating change

Once the design system is off the ground and is being used in real applications, it's imperative to communicate changes, updates, and an ongoing vision to the entire organization.

The tactics for this communication can vary from nuts-and-bolts utilities to more outward-facing marketing efforts. Here are some materials that can help communicate change:

- **Change logs:** "Here's what's changed in the pattern library this month."
- **Roadmap:** "Here's what's coming up over the next few months."
- **Success stories:** "Team X launched this great new application using the design system; read more about how they did it."
- **Tips and tricks:** "Here are a few best practices and considerations for using our system's buttons throughout your application."

Having a base for all these materials is a great idea, and keeping them adjacent to (or even within) the style guide itself makes a lot of sense as well.

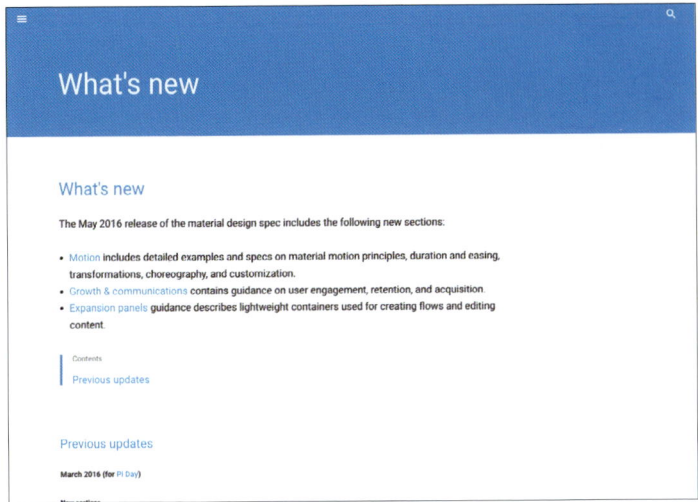

The material design team publishes a handy changelog within its style guide so users can easily learn about the latest updates and improvements to the system.

Design system changes, updates, and requests should be communicated wherever your team hangs out. That may include Slack, Basecamp, GitHub, wikis, Yammer, email lists, company blogs, intranets, and any other internal tools your team uses to communicate and collaborate. If that sounds like a lot of work to you, fear not! Keeping your team and users updated doesn't have to require a huge manual effort. Thanks to the connected nature of our tools, teams can automatically get alerted to changes via software, as Micah Sivitz from Shyp explains:

> Whenever someone makes a pull request, it sends a notification to our #Design slack channel, announcing to the team that there is a proposal change and feedback is required.
>
> - Micah Sivitz, Shyp[95]

[95] https://medium.com/shyp-design/managing-style-guides-at-shyp-c217116c8126

Baking this communication into the team's everyday workflow keeps makers, users, and stakeholders engaged, and helps reassure users that the pattern library is being actively maintained and improved.

Training and support

You wouldn't hand someone a hammer, saw, and screwdriver then say, "All right, you've got what you need; now go and build me a beautiful new house." Knowing how to properly use a tool is often even more important than that tool's availability. Documentation in the form of a style guide is no doubt helpful, but by itself it's not enough. It's essential to provide adequate training and offer ongoing support for your design system's users to ensure they successfully get up and running with the tool kit and continue to create great work with it.

Training users how to work with the design system can take many forms, including:

- **Pair sessions:** Nothing beats pulling up a chair and working together on a project. While more time-intensive than other training vehicles, it's the best way to get makers and users collaborating together, learning how the system works, and exposing new opportunities and shortcomings.
- **Workshops:** From immersive full-day sessions to quick walk-throughs, it's incredibly helpful to set up face-to-face training workshops involving both makers and users. These sessions can help smooth out any misconceptions about the system, help level-up users with hands-on guidance, and create a healthy relationship between the people in charge of maintaining the system and the people in charge of working with it.
- **Webinars:** If workshops or pair sessions aren't possible, or you need to train a lot of users at scale, webinars can be fantastic. Users can tune into online sessions to learn about how to properly use the system. When conducting webinars, be sure to build in plenty of Q&A time to field both audio and typed questions, concerns, and comments.
- **Tutorials:** A series of blog posts and screencasts can neatly encapsulate core concepts of working with the design system. Not only do these help serve as a training tool, but they can serve as a great reference to keep coming back to.

- **Onboarding:** A great way of injecting your design system into your company culture is to bake design system training right into the onboarding process for new employees. New colleagues will understand the importance of modularity, reuse, and all the other benefits a design system brings.

Users will undoubtedly have questions or encounter issues once they get up and running and start building things with the design system. They need to know there's a robust support system in place to help answer any questions, listen to their requirements, and address bugs. There are a host of mechanisms in place to provide support for users, including:

- **Issue trackers:** Tools like JIRA and GitHub Issues are great for users and makers to report bugs and have technical conversations. Users should be aware of the protocol for filing bugs and feel empowered to contribute.
- **Office hours:** Schedule regular times when the design system team is available to field questions, address issues, and talk about what's next for the design system.
- **Slack and chat tools:** The real-time nature of many of our work collaboration tools presents a huge opportunity to keep the pattern-laden conversation going. Thanks to tools like Slack, Yammer, and HipChat, makers and users can engage one another whenever and wherever.
- **Forums:** Communities like Stack Overflow and GitHub have proved to be extremely effective at enabling grassroots, community-driven support. Rather than design system makers becoming a support bottleneck, it can be worthwhile to open up support to the entire user community.
- **Outreach:** Not everyone has the time or the personality to ask questions and suggest changes. Design system makers should be proactive and reach out to developers using the design system to see if they have any issues or concerns. These kinds of actions can help build a genuine and positive relationship between makers and users.

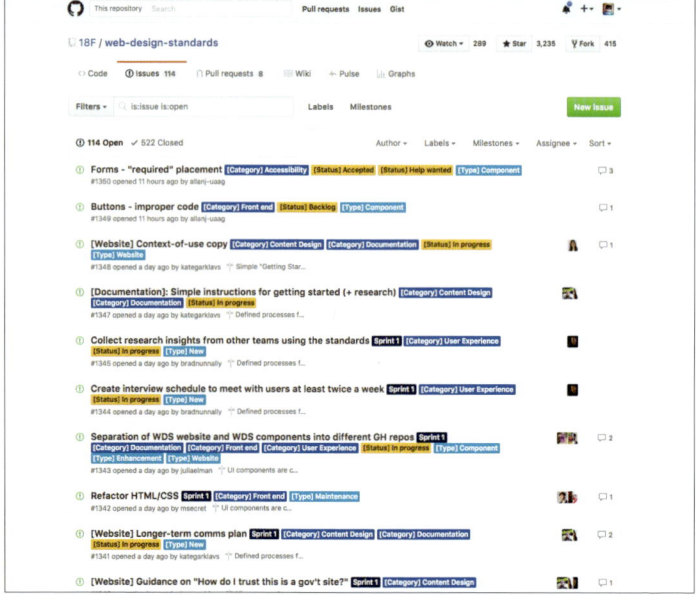

The Draft U.S. Web Digital Standards system tracks issues using GitHub, providing a place for users and makers to file bugs and have conversations about the nitty-gritty.

Thanks to tools like GitHub, design system users don't have to be relegated to the role of dumb consumers. The people who use the system day in and day out can be extremely valuable contributors to the design system if given the chance. Embrace the fact that users are eager to pitch in and make the system as great as it can be. Here are some tactics for encouraging user contributions:

- **Suggestions and pull requests:** Encourage anyone using the design system to suggest changes and new features. Better yet, invite users to submit changes in the form of pull requests that can be merged directly back into the codebase.

- **Individual interviews and roundtable discussions:** It's always a good idea to talk to users, so regularly schedule time to chat with the people who are touching these patterns on a regular basis. Take it all in, listen to both the good and the bad, and collectively determine a plan of attack to address any issues and suggestions.

- **Requests for feedback:** Managing a system that can potentially be deployed to hundreds of applications can be tricky. Before pulling the trigger on decisions that could impact a lot of people, ask for opinions: "We're considering deprecating our carousel pattern and would like to hear what you think."
- **Surveys:** If interviews aren't feasible, you can lean on quick surveys to get a feel for how effective UI patterns and the style guide are. Questions like "On a scale from one to five, how useful is the pattern documentation? Any suggestions?" can help identify blind spots and get users to suggest features that would make their lives easier.
- **Regular "state of the union" meetings:** Schedule regular meetings where the design system team discusses the product roadmap, lessons learned along the way, and suggestions and feedback. Encourage anyone to join the meeting, and be sure to record and distribute these sessions so everyone is aware of the master plan.

Make it public

Communicating change, evangelizing, and setting up proper training and support are all great things to increase your system's visibility. But there's another big opportunity to take your communication strategy to another level: **making your style guide publicly accessible.**

Why? Isn't a style guide merely an internal resource to help people in your organization work better together? What use is it to the outside world? And wouldn't publishing your style guide give away all your trade secrets?

Publishing your style guide for the world to see increases its visibility, increases accountability, and serves as an amazing recruitment tool.

Putting your style guide behind a login or firewall reduces visibility and adds an unnecessary burden to your team and partners, which limits the resource's effectiveness and potential. And the fears about giving away your trade secrets are completely unfounded. These are UI patterns, not nuclear codes.

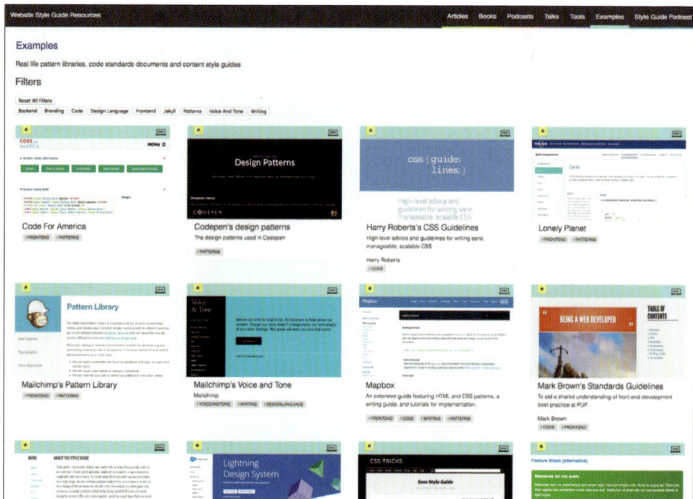

Styleguides.io rounds up over 150 public-facing style guides from organizations across the world.

In addition to making important documentation easier to access, a public style guide **helps create organizational accountability.** Publishing your style guide demonstrates your organization's commitment to the design system, which creates a helpful bit of pressure to keep it an up-to-date and useful resource.

Public-facing style guides are also **hugely helpful for recruiting.** Designers, developers, and people working in other disciplines want to work for organizations that embrace modern digital best practices, and (as we've discussed throughout this book) design systems are quickly becoming an industry-wide best practice. Publishing your style guide sends out a strong Bat-Signal that can attract passionate, pattern-minded people. For instance, style guide expert Jina Bolton went to work at Salesforce after seeing the company's style guide for their Salesforce1 product.

> When I saw [Salesforce's style guide] I thought it was beautiful and it's why I wanted to join this team.
>
> - Jina Bolton[96]

96 http://styleguides.io/podcast/jina-bolton

Since joining Salesforce, she's helped create the ultra-successful Lightning Design System and helps manage their growing design system team. Jina's story is not an isolated one; almost every guest Anna Debenham and I interviewed on the Styleguides Podcast[97] discussed how helpful their public-facing pattern library was for attracting talent. All that means your public style guide makes your organization *more* competitive, not less.

Make it bigger

A visible, cross-disciplinary, approachable pattern library is one that your team will come back to again and again. Use that to your advantage. Since the team's eyeballs are already fixated on that one resource, there's a big opportunity to extend it to include other helpful documentation like the voice and tone, brand, code, design principles and writing guidelines we discussed in chapter 1.

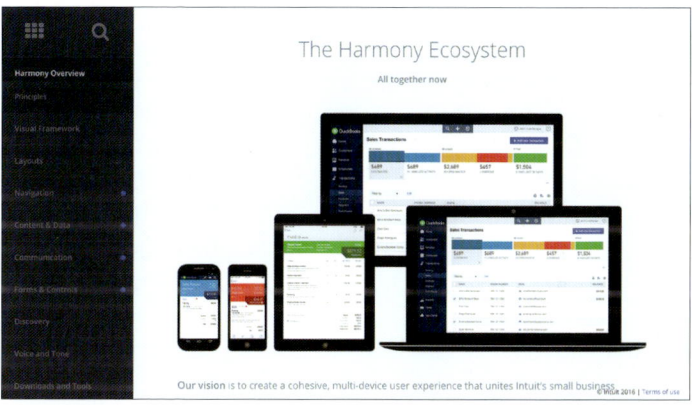

Intuit's Harmony design system includes a pattern library, design principles, voice and tone, marketing guidelines, and more. Housing this helpful documentation under one roof helps increase its visibility and effectiveness.

Now, your organization may not need to implement every flavor of style guide out there, but the point is that **creating a centralized style guide hub builds more awareness of best practices, increasing the documentation's effectiveness.**

97 http://styleguides.io/podcast

Another way to extend the functionality of the pattern library is to include guidelines for native platform patterns alongside web-based patterns. We can look to Intuit's Harmony design system once again for an example of how native mobile platform patterns for iOS and Android can live beside their web-based counterparts.

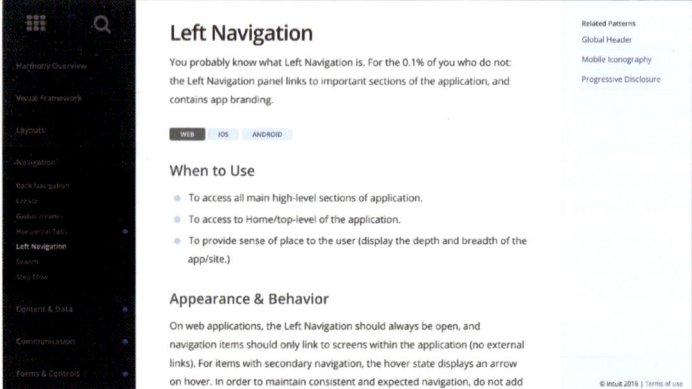

Intuit's Harmony pattern library includes buttons to switch between web, iOS, and Android for each pattern. This allows the team to maintain a mostly consistent design system across platforms but also document pattern divergences when they occur.

Make it context-agnostic

The way your UI patterns are named will undoubtedly shape how they are used. **The more agnostic pattern names are, the more versatile and reusable they become.**

Because we tend to establish UI patterns in the context of a broader page, it can be tempting to name components based on where they live. But rather than naming your component "homepage carousel" (forgive my morbid obsession with carousels), you can simply call it "carousel," which means you can now put carousels everywhere! (But for the love of all that is holy, please don't.)

Another challenge for naming *display* patterns is that we tend to get distracted by the *content* patterns that live inside them. For instance, if working on an e-commerce site, you may be tempted to call a block containing a product image and title a "product card."

But naming things in this manner immediately limits what type of content can live inside it. By naming the pattern simply "card," you can put all sorts of content patterns inside it: products, promotions, store locations, and so on.

Fair warning: **naming things is really freaking hard.** But there are strategies to help you create robust names for your patterns. Conducting an interface inventory (as detailed in chapter 4) helps remove patterns from the context of the page where they normally reside, meaning your team can create names that aren't distracted by their context. I've conducted naming exercises with teams where we've blurred out the content residing inside a pattern so everyone can focus on the pattern's *structure* rather than the *content* that lives inside it.

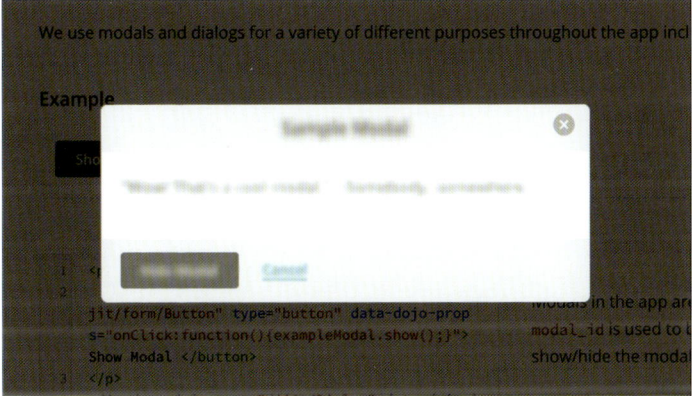

A good exercise when naming patterns is to blur out the content so your names reflect the patterns' structures rather than the content living inside them.

While naming things will always be a challenge, pattern names that are agnostic to context and content will be more portable, reusable, and versatile.

Make it contextual

Showcasing UI patterns in a pattern library is all well and good, but **you need to demonstrate context for design system users to understand how and where to properly use them.** Most pattern libraries show a demo of each UI pattern, but as we've discussed, those patterns don't live in a vacuum. Where exactly are these patterns used?

One way to demonstrate context might include showing screenshots or videos of a component in action. Material design's documentation does a fantastic job at this; each component is rich with photos, videos, and usage details to give users a clear understanding of what these patterns look like in the context of an application, and demonstrate how each pattern should be used.

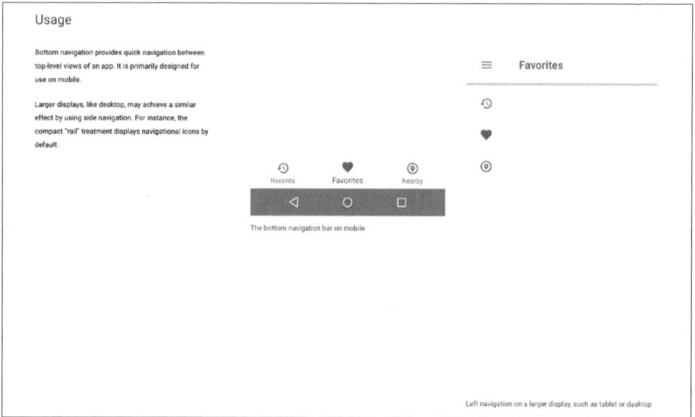

Material design's component library doesn't just contain an example of each component; it thoroughly documents the component's usage with plenty of images and videos to support it.

Another way to show context is to provide lineage information for each pattern. As we discussed in Chapter 3, a tool like Pattern Lab automatically generates this information, letting you see which patterns make up any given component in addition to showing where each component is employed. This provides a sort of pattern paper trail that helps immensely with QA efforts, as it highlights exactly which patterns and templates would need to be tested if changes were made to a particular pattern.

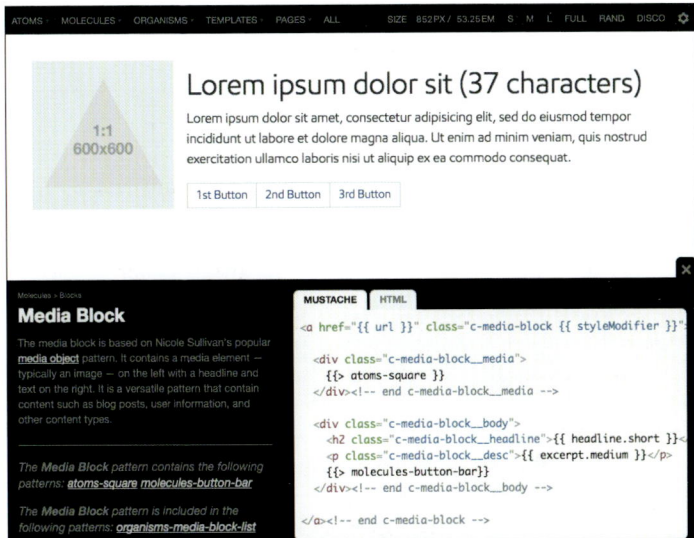

Tools like Pattern Lab provide lineage information, allowing teams to see which smaller components are included in any given component, as well as where each pattern gets used.

Make it last

Making a design system is an incredibly and important endeavor. But without proper maintenance, the value of your design system will depreciate much like a car that's just been driven off the dealer's lot. Instead, your design system should be like a bottle of fine wine that increases in value over time.

With proper maintenance, your design system should increase in value over time like a bottle of fine wine, rather than a used car that's just been driven off the lot. Image credit: Sabin Paul Croce on Flickr[98] and Ray Larabie on Flickr[99]

As we've discussed throughout this chapter, making your design system stand the test of time requires a significant amount of time and effort. But isn't that the case with all living things? Animals need to eat, and plants need water and sunlight in order to survive. Creating a living design system means giving it attention and care in order for it to continue to thrive.

All that effort not only creates a better present for your organization, but sets you up for long-term success. Establishing a clear governance plan, communicating change, and implementing the other advice found in this chapter helps the design system take root and become an integral part of your organization's workflow. Creating the damn thing in the first place is the hard part, but once established, you have a solid foundation with which to build on for years to come. Even if you were to burn everything down and rebuild a new system from the ground up, you'll find your UIs will still need buttons, form fields, tabs, and other existing components. And you'll need a happy home to display and document the system. Don't throw the baby out with the bathwater!

98   https://flic.kr/p/on4ffK
99   https://flic.kr/p/e35AtD

So there you have it. To make a maintainable design system, you should:

- **Make it official** by allocating real time, money, and resources to your design system.
- **Make it adaptable** by counting on change and establishing a clear governance plan.
- **Make it maintainable** by seeking the holy grail and making it easy to deploy and communicate changes to the design system.
- **Make it cross-disciplinary** by making your pattern library a watering hole the entire organization can gather around.
- **Make it approachable** by making an attractive, easy-to-use style guide with helpful accompanying documentation.
- **Make it visible** by communicating change, evangelizing the design system, and making it public.
- **Make it bigger** by including brand, voice and tone, code, design principles, and writing guidelines.
- **Make it agnostic** by naming patterns according to their structure rather than their context or content.
- **Make it contextual** by demonstrating what patterns make up a particular pattern and showing where each pattern is used.
- **Make it last** by laying a solid foundation with which to build on for years to come.

Go forth and be atomic

We're tasked with making a whole slew of products, sites, and applications work and look great across a dizzying array of different devices, screen sizes, form factors, and environments. I hope that the concepts covered in this book give you solid ground to stand on as you bravely tackle this increasingly diverse digital landscape. By creating design systems, being deliberate in how you construct user interfaces, establishing a collaborative and pattern-driven workflow, and setting up processes to successfully maintain your design system, I hope you and your team can create great things together. Go forth and be atomic!

Thanks & Acknowledgements

This book is dedicated to my amazing wife **Melissa**, who supports all of my crazy ideas and somehow puts up with all of my shit. Thank you. I love you.

I'd like to give a massive thank you to **Dave Olsen**, who took my nascent, poorly programmed Pattern Lab[100] concept and transformed it into a legit and amazing piece of software. Thanks to the tireless work of Dave and **Brian Muenzenmeyer**, Pattern Lab is helping teams all over the world create atomic design systems. I'm forever grateful for all of your superb work, and consider myself fortunate to call you friends.

To **Josh Clark** and **Dan Mall** for helping solidify atomic design as a methodology and for writing the foreword to this book. You trusted me enough to run with this approach and somehow convinced our clients it wasn't totally insane. Without your input and the crazy-smart brains of early collaborators like **Jennifer Brook**, **Jonathan Stark**, **Robert Gorrell**, **Kelly Shaver**, and **Melissa Frost**, this book wouldn't have existed.

Thanks to **Owen Gregory** for copyediting the book's manuscript and taking on the Herculean task of making me sound reasonably coherent. Thanks to **Rachel Andrew** for wrangling all the technical stuff that goes into making ebooks. And a big thanks to **Rachel Arnold Sager** for all your work getting the print version of the book laid out and ready for the printer.

To **Anna Debenham** for all your amazing thinking about front-end style guides, your book[101] on the topic, and your willingness to co-host a podcast all about style guides with me. I'm proud of the work we've done on Styleguides.io[102] and I'm so happy we got to work together.

To **Jonathan Snook** for your fantastic SMACSS methodology, and for taking the time to guide me through the process of self-publishing a book. Thanks for making such a scary endeavor a lot more approachable.

---

100 http://patternlab.io/
101 http://maban.co.uk/projects/front-end-style-guides/
102 http://styleguides.io/

To **Stephen Hay**, who was the first person I heard articulate the need to break interfaces into smaller pieces. Thanks for being a continued source of wisdom and sarcasm.

To **Andy Clarke**, who was talking about design systems and atoms[103] before it was the hip thing to do. Thank you for all your writing and thinking, but you're still not getting my dog.

To Dave "Tiny Bootstraps" Rupert, Susan Robertson, Samantha Warren, Jina Bolton, Nathan Curtis, Paul Robert Lloyd, Harry Roberts, Nicole Sullivan, Brett Jankord, Tyler Sticka, Lincoln Mongillo, Nicholas Gallagher and the many others who have advanced the concepts of design systems, pattern libraries, and style guides. Thanks for helping me and so many others think more modularly.

To Jeffrey Zeldman, Eric Meyer, Marc Thiele, Vitaly Friedman, and all the other conference organizers who gave me the opportunity to stumble onto the stage to ramble on about the concepts contained in this book.

This book would not have been possible if it weren't for the amazing work done by some amazing people in the web community. I'm so incredibly fortunate to work in such an open, sharing, and collaborative community; every day I look forward to learning new things from you all.

And last, but certainly not least, thanks so much to my family for all your love and amazing support over the years.

103 http://stuffandnonsense.co.uk/blog/about/an-extract-from-designing-atoms-and-elements

Resources

Chapter 1

- Scope Components, Not Pages
  http://bradfrost.com/blog/post/scope-components-not-pages/
- W3C Principles of Design - Modular Design
  http://www.w3.org/DesignIssues/Principles.html#Modular
- YUI Library
  http://yuilibrary.com/
- jQuery U
  http://jqueryui.com/
- Future Friendly Manifesto
  http://futurefriendlyweb.com/
- Riding the Magic Escalator of Acquired Knowledge
  http://www.uie.com/articles/magic_escalator/
- Agile Manifesto
  http://www.agilemanifesto.org/
- Scrum software development
  http://en.wikipedia.org/wiki/Scrum_%28software_development%29
- Lean software development
  http://en.wikipedia.org/wiki/Lean_software_development
- Principles behind the Agile Manifesto
  http://www.agilemanifesto.org/principles.html
- DIY Process
  http://cognition.happycog.com/article/diy-process
- For a Future Friendly Web
  http://bradfrost.com/blog/web/for-a-future-friendly-web/
- Adapting Ourselves to Adaptive Content
  http://karenmcgrane.com/2012/09/04/adapting-ourselves-to-adaptive-content-video-slides-and-transcript-oh-my/
- COPE: Create Once, Publish Everywhere
  http://www.programmableweb.com/news/cope-create-once-publish-everywhere/2009/10/13

- Learning JavaScript Design Patterns
  http://addyosmani.com/resources/essentialjsdesignpatterns/book/
- OOCSS
  http://oocss.org/
- SMACSS
  https://smacss.com/
- MindBEMding – getting your head 'round BEM syntax
  http://csswizardry.com/2013/01/mindbemding-getting-your-head-round-bem-syntax
- An extract from Designing Atoms and Elements
  http://stuffandnonsense.co.uk/blog/about/an-extract-from-designing-atoms-and-elements
- Style Tiles
  http://styletil.es/
- Element Collages
  http://danielmall.com/articles/rif-element-collages/
- BDConf: Stephen Hay presents Responsive Design Workflow
  http://bradfrost.com/blog/mobile/bdconf-stephen-hay-presents-responsive-design-workflow/
- Responsive Web Design
  http://alistapart.com/article/responsive-web-design
- This Is Responsive
  http://bradfrost.github.io/this-is-responsive/index.html
- Foundation by ZURB
  http://foundation.zurb.com/
- Bootstrap
  http://getbootstrap.com/
- Github
  https://github.com/
- Responsive Deliverables
  http://daverupert.com/2013/04/responsive-deliverables/

- Style Guides
  http://bradfrost.com/blog/post/style-guides/
- Material Design
  http://www.google.com/design/spec/material-design/introduction.html
- Voice and Tone: Creating content for humans
  http://www.slideshare.net/katekiefer/kkl-c-sforum
- Voice & Tone
  http://voiceandtone.com/
- Writing for the Web
  http://www.dal.ca/webteam/web_style_guide/writing_for_the_web.html
- Front End Style Guides
  http://maban.co.uk/projects/front-end-style-guides/
- Style Guides Podcast with Federico Holgado
  http://styleguides.io/podcast/federico-holgado/
- Dennis Crowley: "The Hard Part Is Building The Machine That Builds The Product"
  http://techcrunch.com/2011/03/03/founder-stories-foursquare-crowley-machine/
- Style Guide Examples on Styleguides.io
  http://styleguides.io/examples.html

Chapter 2

- Systemic Design
  http://us5.campaign-archive1.com/?u=7e093c5cf4&id=ead8a72012&e=ecb25a3f93
- Josh Duck's Periodic Table of HTML Elements
  http://smm.zoomquiet.io/data/20110511083224/index.html
- HTML element reference
  https://developer.mozilla.org/en-US/docs/Web/HTML/Element
- Single responsibility principle
  https://en.wikipedia.org/wiki/Single_responsibility_principle
- Structure First. Content Always.
  http://www.markboulton.co.uk/journal/structure-first-content-always

- The Shape of Design
  http://read.shapeofdesignbook.com/chapter01.html
- GE's Predix Design System
  https://medium.com/ge-design/ges-predix-design-system-8236d47b0891#.uo68yjo9g

Chapter 3

- Pattern Lab
  http://patternlab.io
- Dave Olsen
  http://dmolsen.com
- Brian Muenzenmeyer
  http://www.brianmuenzenmeyer.com/
- Pattern Lab's documentation
  http://patternlab.io/docs/
- Don't repeat yourself
  https://en.wikipedia.org/wiki/Don't_repeat_yourself
- Mustache
  https://mustache.github.io/
- Pattern Lab's pseudo-patterns
  http://patternlab.io/docs/pattern-pseudo-patterns.html
- Responsive Web Design
  http://alistapart.com/article/responsive-web-design
- Container Queries: Once More Unto the Breach
  http://alistapart.com/article/container-queries-once-more-unto-the-breach
- Ish
  http://bradfrost.com/demo/ish/
- Website style guide resources examples
  http://styleguides.io/examples.html
- Rizzo style guide
  http://rizzo.lonelyplanet.com/
- Website style guide resources tools
  http://styleguides.io/tools.html

Chapter 4

- Content inventory
  https://en.wikipedia.org/wiki/Content_inventory
- Google Slides interface inventory template
  https://docs.google.com/presentation/d/1GqFmiDV_NqKi36fXAwD3WTJL5-JV-gHL7XVD2fVeL0M/edit?usp=sharing
- The Media Object Saves Hundreds of Lines of Code
  http://www.stubbornella.org/content/2010/06/25/the-media-object-saves-hundreds-of-lines-of-code/
- Surfacing Invisible Elements
  http://bradfrost.com/blog/post/surfacing-invisible-elements/
- CSS Stats
  http://cssstats.com/
- Stylify Me
  http://stylifyme.com/
- Multiscreen UX Design
  http://store.elsevier.com/Multiscreen-UX-Design/Wolfram-Nagel/isbn-9780128027295/
- The Post-PSD Era
  http://danielmall.com/articles/the-post-psd-era/
- Consensual hallucination
  https://adactio.com/journal/4443
- Waterfall model
  https://en.wikipedia.org/wiki/Waterfall_model
- Development Is Design
  http://bradfrost.com/blog/post/development-is-design
- Time to stop showing clients static design visuals
  https://stuffandnonsense.co.uk/blog/about/time_to_stop_showing_clients_static_design_visuals
- Mobile First
  http://www.lukew.com/ff/entry.asp?933
- Content & Display Patterns
  http://danielmall.com/articles/content-display-patterns/
- Jennifer Brook
  http://jenniferbrook.co/about

- The 20 Second "Gut" Test
  http://goodkickoffmeetings.com/2010/04/the-20-second-gut-test/
- Style Tiles
  http://styletil.es/
- Style prototype
  http://sparkbox.github.io/style-prototype/
- Element Collages
  http://danielmall.com/articles/rif-element-collages/
- Piles of Ideas
  http://jasonsantamaria.com/articles/piles-of-ideas
- CodePen
  http://codepen.io/
- Dan Mall on The Pastry Box Project
  https://the-pastry-box-project.net/dan-mall/2012-september-12

Chapter 5

- Nathan Curtis on Twitter
  https://twitter.com/nathanacurtis/status/656829204235972608
- A Design System isn't a Project. It's a Product, Serving Products.
  https://medium.com/eightshapes-llc/a-design-system-isn-t-a-project-it-s-a-product-serving-products-74dcfffef935#.4umtnfxsx
- The Salesforce Team Model for Scaling a Design System
  https://medium.com/salesforce-ux/the-salesforce-team-model-for-scaling-a-design-system-d89c2a2d404b
- Vanilla framework
  http://ubuntudesign.github.io/vanilla-framework/
- Getting Vanilla ready for v1: the roadmap
  http://design.canonical.com/2016/07/getting-vanilla-ready-for-v1-the-roadmap/
- Sass Deprecate
  https://github.com/salesforce-ux/sass-deprecate
- The Way We Build
  http://airbnb.design/the-way-we-build/

- Rizzo style guide
  http://rizzo.lonelyplanet.com/
- Chasing the Holy Grail
  https://medium.com/@marcelosomers/chasing-the-holy-grail-bbc0b7cce365#.ay1xeej7d
- Drupal
  https://www.drupal.org
- Twig
  http://twig.sensiolabs.org
- Introducing Pattern Lab Starter 8
  https://www.phase2technology.com/blog/introducing-pattern-lab-starter-8/
- Draft U.S. Web Digital Standards
  https://standards.usa.gov/
- Managing Style Guides at Shyp
  https://medium.com/shyp-design/managing-style-guides-at-shyp-c217116c8126
- Style Guides with Jina Bolton
  http://styleguides.io/podcast/jina-bolton
- Website style guide resources podcasts
  http://styleguides.io/podcast

About the Author

Brad Frost[104] is a web designer, speaker, consultant, and musician located in beautiful Pittsburgh, PA. He's passionate about creating web experiences that look and function beautifully on a never-ending stream of connected devices, and loves helping others do the same. He's helped create several tools and resources for web designers, including Pattern Lab[105] (with Dave Olsen and Brian Muenzenmeyer), Styleguides.io[106] (with Anna Debenham), This Is Responsive[107], WTF Mobile Web[108] (with Jen Simmons), and Mobile Web Best Practices.

104 http://bradfrost.com/
105 http://patternlab.io/
106 http://styleguides.io/
107 https://bradfrost.github.io/this-is-responsive/
108 http://wtfmobileweb.com/